Vitamin Tolerance of Animals

Subcommittee on Vitamin Tolerance
Committee on Animal Nutrition
Board on Agriculture
National Research Council

NATIONAL ACADEMY PRESS
Washington, D.C.   1987

National Academy Press   2101 Constitution Avenue, NW   Washington, DC 20418

NOTICE: The project that is the subject of this report was approved by the Governing Board of the National Research Council, whose members are drawn from the councils of the National Academy of Sciences, the National Academy of Engineering, and the Institute of Medicine. The members of the committee responsible for the report were chosen for their special competences and with regard for appropriate balance.

This report has been reviewed by a group other than the authors according to procedures approved by a Report Review Committee consisting of members of the National Academy of Sciences, the National Academy of Engineering, and the Institute of Medicine.

The National Research Council was established by the National Academy of Sciences in 1916 to associate the broad community of science and technology with the Academy's purposes of furthering knowledge and of advising the federal government. The Council operates in accordance with general policies determined by the Academy under the authority of its congressional charter of 1863, which establishes the Academy as a private, nonprofit, self-governing membership corporation. The Council has become the principal operating agency of both the National Academy of Sciences and the National Academy of Engineering in the conduct of their services to the government, the public, and the scientific and engineering communities. It is administered jointly by both Academies and the Institute of Medicine. The National Academy of Engineering and the Institute of Medicine were established in 1964 and 1970, respectively, under the charter of the National Academy of Sciences.

This study was supported by the U.S. Department of Agriculture, Agricultural Research Service, under Agreement No. 59-32U4-5-6, and by the Center for Veterinary Medicine, Food and Drug Administration of the U.S. Department of Health and Human Services, under Cooperative Agreement No. FD-U-000006-06-1. Additional support was provided by the American Feed Industry Association, Inc. Any opinions, findings, conclusions, or recommendations expressed in this publication are those of the authoring subcommittee and do not necessarily reflect the views of the sponsors.

Library of Congress Cataloging-in-Publication Data

National Research Council (U.S.). Subcommittee on
  Vitamin Tolerance.
  Vitamin tolerance of animals.

  Bibliography: p.
  Includes index.
  1. Vitamin tolerance in animals. I. Title.
  SF98.V5N37 1986   636.08'52   86-28466
ISBN 0-309-03728-X

Printed in the United States of America

Preface

Optimal animal health and productivity are achieved by providing an animal with the correct amount and form of each essential nutrient. Good nutritional practice requires supplementing practical diets with vitamin levels that exceed the bare minimum needed to prevent deficiency diseases. Yet, vitamins added to supplement feeds, which may have lost vitamins during processing and storage, or to protect the animal from stress or stimulate its immune system may necessitate the use of much higher levels. The goal in animal feeding is to supplement diets with vitamin levels that are adequate for nutritional needs and accommodate the practical conditions of feed manufacture and storage.

To address these concerns, the Board on Agriculture's Committee on Animal Nutrition appointed the Subcommittee on Vitamin Tolerance in 1983. The subcommittee's report is the most comprehensive summary of current data about the vitamin tolerances of animals. It will be helpful to livestock producers, livestock extension specialists, animal nutritionists, animal nutrition students, and others interested in the subject.

The presumed upper safe levels of vitamins in this book are meant to be used as guidelines to ensure that vitamin supplementation does not adversely affect animal health. The subcommittee's objectives were to focus on: (1) vitamin tolerance of domestic and laboratory animals under different nutritional and physiological states, (2) biological measures that can be used as criteria to establish tolerance, and (3) areas of incomplete knowledge.

The first two objectives were, unfortunately, impossible to fulfill completely. For most of the vitamins, information in the scientific literature on vitamin toxicities was incomplete with respect to the range of species studied and the quantitative aspects, such as dose-response definitions. In some cases, the literature indicated rather clearly the ranges of intake that would produce certain adverse responses in certain species; in others, it indicated little evidence of toxicity.

The third objective is significant. The subcommittee's indicated values for the presumed safe levels are limited in scope by gaps in current information. Areas lacking adequate information are identified. It is the subcommittee's hope that this report points to the limits of current knowledge and important areas of research that can contribute to improvements in animal nutrition.

The following individuals were responsible for respective sections of the report: Robert Blair, riboflavin, vitamin B_6 (pyridoxine), biotin, vitamin B_{12}, and choline; Gerald F. Combs, Jr., vitamin E and niacin; John W. Hilton, ascorbic acid and thiamin; Ronald L. Horst, vitamin D; George E. Mitchell, Jr., vitamin A; and John W. Suttie, vitamin K, pantothenic acid, and folic acid.

The report was reviewed by the Committee on Animal Nutrition, the Board on Agriculture, and outside reviewers. The subcommittee is grateful for the efforts of these individuals. We also thank Annette Bates, Esther Collins, Karen Davis, Andra Hinds, Frances Newsome, and Pamela Senter, at our respective institutions, for their administrative assistance. The subcommittee especially acknowledges the contributions of Selma Baron, who served as staff officer during the early preparation of this report, and assistant editor Grace Jones Robbins, who helped us complete our task.

GERALD F. COMBS, JR.
Chairman
Subcommittee on
Vitamin Tolerance

Contents

Figures and Tables

FIGURES

TABLES

Introduction

Although vitamins are essential in the diet for normal health, they may have adverse physiological effects when consumed in excessive amounts. Vitamin intake levels that are above those required to prevent specific deficiency syndromes but still below levels shown to produce toxicity can be said to be tolerated. In this sense, tolerance indicates the absence of deleterious effects of vitamin intakes above those needed to prevent nutritional deficiency disorders. Understanding of the range of tolerance for each vitamin is important in the safe and adequate feeding of domestic and laboratory animals. To ensure optimal animal health, vitamin intakes must be maintained within these ranges, whether those intakes are through the use of vitamin supplements to animal feeds or vitamins in enteral preparations.

The term vitamin is generally accepted to describe an organic compound that is (1) a component of a natural food, but is distinct from carbohydrate, fat, protein, and water; (2) present in most foods in minute amounts; (3) essential for normal metabolism in animals and consequently required for normal health and physiological functions such as growth, development, maintenance, and reproduction; (4) a cause of a specific deficiency disease or syndrome if absent from the diet or improperly absorbed or utilized; and (5) unable to be synthesized by the host in sufficient amounts to meet physiological demands and therefore must be obtained from the diet. Most of the traditional vitamins, such as the vitamins A, E, K, B_6, and B_{12}, thiamin, riboflavin, folic acid, pantothenic acid, and biotin, satisfy each of these criteria.

Vitamins C and D, niacin, and choline, however, are considered vitamins only in certain contexts. For example, ascorbic acid is produced metabolically by most species from glucose via the uronic acid pathway. Only the few species that lack the enzyme L-gulonolactone oxidase (for example fishes, the guinea pig, the Indian fruit bat, the red-vented bulbul, and higher primates) require preformed ascorbic acid in the diet. For them, ascorbic acid is vitamin C. Vitamin D, while biosynthesized by animals upon exposure to the ultraviolet radiations in sunlight, is also a vitamin in the context of the environmentally controlled housing (and the associated use of artificial lighting) commonly used in modern systems of confined rearing of animals. Niacin is produced metabolically from the indispensible amino acid tryptophan. Nevertheless, the efficiency of this conversion is low enough that most animals (especially cats, fishes, and ducks) fed the tryptophan levels typical of mixed practical diets also need a dietary source of niacin. Choline, which is produced metabolically from the amination and subsequent methylation of serine, appears not to be biosynthesized in sufficient amounts to satisfy the metabolic demands of rapidly growing chicks and poults. In the context of young poultry, therefore, choline is a vitamin. To summarize, compounds that satisfy the definition of a vitamin noted above are traditionally considered vitamins and are examined in this report.

As pointed out above, the concept of vitamin tolerance indicates exposure to vitamin levels that prevent deficiency diseases but produce no signs of intoxication. Thus, it must be the goal of the animal producer, feed manufacturer, and veterinarian to provide vitamins to animals at levels that are sufficient to prevent vitamin deficiency disease and that can be well tolerated. To ensure vitamin tolerance in animal production, there is the practical need to define maximum tolerable levels. The current scientific literature, however, is not complete enough to support the estimation of maximum tolerable levels for any vitamin. In most cases, however, one can use the literature to estimate ranges of vitamin intake that can be presumed to be safe. For the purposes of this report, these upper levels of vitamin intake are

1

called presumed upper safe levels. It is reasonably clear that these levels will not impair any aspect of animal health or production efficiency and will not accumulate as hazardous residues in human food products derived from animals. Presumed upper safe levels for specific vitamins, therefore, will be less than maximum tolerable levels. The presumed upper safe level for each vitamin will vary according to the chemical form of the vitamin; the route of administration, for example, oral versus parenteral; the length of exposure, for example, acute versus chronic; and the particular species and developmental state of the animal, according to such species-dependent variables as degree of endogenous synthesis and efficiencies of absorption, retention, and tissue storage.

In order to identify presumed upper safe levels of each of the traditional vitamins, the pertinent scientific literature is presented, and the major findings are discussed. In a few cases, the presumed upper safe levels for certain vitamins with certain species are identified through a straightforward review of published information. In most cases, however, the current scientific literature is not complete enough to indicate these levels unequivocally. For these vitamins, presumed upper safe levels are estimated by extrapolation and inference from the available data. The focus of this report is on vitamin intake by domestic and laboratory animal species with emphasis on dietary exposure. Where information from studies with humans is enlightening, it has been included in considerations of vitamin tolerance of animals.

When planning animal diets, it is good practice to consider factors such as vitamin bioavailability and stability during storage and animal physical and metabolic stresses. These factors can influence the levels of vitamins and other nutrients that animals require. Thus, vitamin levels used in practical animal feeding are expected to be greater than those traditionally recognized as the nutritional requirements. The use of such "margins of safety," however, carries with it risks of vitamin toxicity. In an attempt to help minimize risk, the presumed upper safe levels of vitamins have been identified in this report.

Vitamin A

Vitamin A was the first fat-soluble vitamin to be discovered and characterized. It has essential roles in vision, bone and muscle growth, reproduction, and maintenance of healthy epithelial tissue. Either vitamin A or a precursor must be provided in the diet. However, it is among the most highly variable nutrients in feeds. Plants do not contain vitamin A, and most grains other than yellow corn are almost devoid of the carotenoid precursors that provide plant sources of vitamin A activity. The concentrations of carotenoids in the vegetative portions of plants vary widely according to geographic location, maturity, method of harvest, amount and type of processing, length and conditions of storage, and exposure to high temperature, sunlight, and air. Eggs and selected poultry, fish, animal products (especially liver, milk, and milk products), and fats may contain high levels of vitamin A or carotene, but these levels reflect vitamin A or carotenoids present in the diets of those animals. Consequently, vitamin A is a frequent nutritional concern, which has been extensively reviewed (Moore, 1957; Mitchell, 1967; Eaton, 1969; Olson, 1969, 1984; Ullrey, 1972; Bauernfeind et al., 1974; Goodman, 1980).

Most workers rank vitamin A deficiency next to protein and calorie deficiency as a worldwide health problem. It is the most important vitamin in ruminant animal diets and is almost universally added to commercial diets for nonruminant animals. Vitamin A toxicity due to the consumption of rich natural sources such as polar bear's liver and fish oils is well documented (Pitt, 1985) in humans and laboratory animals but is apparently rare in domestic animals. The potential for nutritional abuse leading to toxicity has been increased by the availability of economical sources of synthetic vitamin A, however. Pharmacological use of retinoids to treat skin disease (Moore, 1957) and cancer (Ong and Chytil, 1983) requires levels that make toxicity a major hazard.

NUTRITIONAL ROLE

Dietary Requirements of Various Species

Vitamin A is an essential nutrient for all species of mammals, birds, and fishes studied and is also essential in many lower forms of life. The dietary requirements for most adequately studied species are between 1,500 and 4,000 IU/kg of diet. (One IU provides the vitamin A activity of 0.3 μg all-*trans*-retinol.) Based on limited data, requirements for Japanese quail have been set at 5,000 IU/kg of diet (National Research Council, 1984b) and those for cats (National Research Council, 1978a), nonhuman primates (National Research Council, 1978d), and some warmwater fishes (National Research Council, 1983) at 10,000 IU/kg of diet. Inadequate vitamin A intake may result in reduced feed intake, edema, lacrimation, xeropthalmia, nyctalopia (night blindness), slow growth, low conception rates, abortion, stillbirths, blindness at birth, abnormal semen, reduced libido, susceptibility to respiratory and other infections, and death. Only nyctalopia has been proven unique to vitamin A deficiency. When several of these other signs are present, vitamin A deficiency should be suspected. It may be verified by ophthalmoscopic examination, liver biopsy and assay for near absence of vitamin A (retinyl esters), blood assay for vitamin A (concentrations of retinol below 20 μg/100 ml are considered below normal in most species), spinal fluid pressure testing for an above-normal elevation, conjunctival smear examination for epithelial keratinization, and response to vitamin A therapy.

Biochemical Functions

The classic work of Wald (1968) has defined the biochemical role of vitamin A in night vision. Key steps in

3

this process are oxidation of retinol to retinal and isomerization of the *trans* form to 11-*cis*-retinal, which combines with the protein opsin to form rhodopsin, which is known as visual purple. The 11-*cis*-3-dehydroretinal form of naturally occurring vitamin A_2 is active in fish but not in mammals or birds. The molecular bases for the roles of vitamin A in growth, reproduction, and epithelial health have been studied extensively but remain incompletely understood. The most widely accepted hypotheses propose a role in synthesis of glycoproteins that may control cell differentiation and involvement in the control of gene expression (Olson, 1984).

FORMS OF THE VITAMIN

Vitamin A activity is a generic term for β-ionone derivatives having the biological activity of all-*trans*-retinol. In plants this activity is present only in the form of carotenoid precursors of all-*trans*-retinol. The most active of these precursors is β-carotene, which can be cleaved by intestinal enzymes to yield two moles of all-*trans*-retinol per mole of β-carotene. Foodstuffs of animal origin may contain either carotenoids or retinoids. The most significant retinoids in animal metabolism are the alcohol (all-*trans*-retinol), the aldehyde (11-*cis*-retinal and 11-*cis*-3-dehydroretinal), and the acid (all-*trans*-retinoic acid) forms, as well as retinyl esters (especially retinyl palmitate) and retinyl β-glucuronide. Structural formulas for most of these are given in Figure 1.

ABSORPTION AND METABOLISM

Various forms of vitamin A and carotenoids are absorbed mainly in conjunction with lipids. (See Table 1 for the relative vitamin A activity of carotenoids.) Carotenoids are normally converted to retinol in the intestinal mucosa but may also be converted in the liver and other organs, especially in yellow fat species such as cattle and poultry. Either dietary retinol or retinol resulting from conversion of carotenoids is then esterified with a long-chain fatty acid, usually palmitate. Dietary retinyl esters are hydrolyzed to retinol in the intestine; they are absorbed as the free alcohol and then re-esterified in the mucosa. In mammals, the retinyl esters are transported mainly in association with lymph chylomicrons to the liver where they are hydrolyzed to retinol and re-esterified for storage. Hydrolysis of the ester storage form mobilizes vitamin A from the liver as free retinol. Retinol is released from the hepatocyte as a complex

FIGURE 1 Major compounds of the vitamin A group.

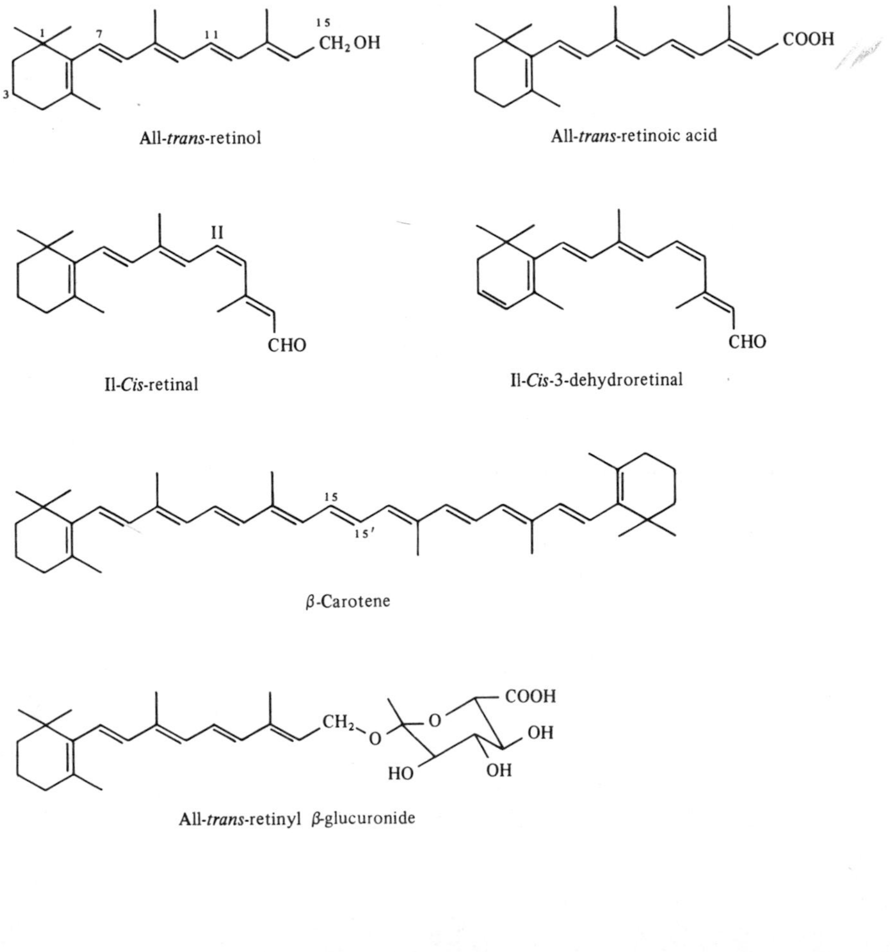

All-*trans*-retinol

All-*trans*-retinoic acid

Il-*Cis*-retinal

Il-*Cis*-3-dehydroretinal

β-Carotene

All-*trans*-retinyl β-glucuronide

TABLE 1 Relative Vitamin A Activity of Carotenoids

Compound	Relative Biological Activity[a]
Retinol (all-*trans*)	100
Natural or artificial esters of all-*trans*-retinol	100
Retinal (all-*trans*)	100
Cis-isomers of retinol	23–75
Phenyl or methyl esters of retinol	10–100
Vitamin A₂	30
β-Carotene	50
α-Carotene	26
ν-Carotene	21
Cryptoxanthin	28
Zeaxanthin	0

[a]In reference to all-*trans*-retinol set at 100. Comparisons are on a molar basis for retinoids but on a weight basis for comparisons of carotenoids with retinol.

SOURCE: Derived from Moore (1957) and Olson and Lakshmanan (1969) from data for chicks and rats. Vitamin A₂ data is based on liver storage by fish.

with retinol-binding protein; it is transported in this form to the tissues. The main excretory pathway is by elimination as glucuronide conjugates in the bile. Glucuronide formation may follow irreversible oxidation to retinoic acid. Retinoic acid supports growth and cell differentiation but not the functions of vitamin A in vision and reproduction. The enterohepatic circulation may provide an important means of conserving vitamin A prior to fecal excretion. Small amounts of glucuronide and chain-shortened metabolites may be excreted in the urine.

HYPERVITAMINOSIS

A voluminous amount of literature clearly indicates that vitamin A has the potential to act as a cumulative toxicant in most species that have been studied (Nieman and Obbink, 1954; Moore, 1957; Hayes and Hegsted, 1973; Bauernfeind, 1980; Agricultural Research Council, 1980, 1981; Ong and Chytil, 1983; Olson, 1984).

Table 2 summarizes many of the published reports. Acute single dose toxicity has been well-documented in humans (Nieman and Obbink, 1954; Hayes and Hegsted, 1973). Massive doses elicit responses within hours. Reactions may include general malaise, anorexia, nausea, hyperirritability, peeling skin, muscular weakness, twitching, convulsions, paralysis, and death. If death is avoided, recovery from these signs of toxicity is usually prompt upon removal of vitamin A from the diet.

Chronic toxicity typically results from intakes 100 to 1,000 times nutritional requirements for a prolonged period but has been observed at intakes of approximately 10 times the specific requirement (Olson, 1984). The most characteristic signs of hypervitaminosis A are skeletal malformation, spontaneous fractures, and internal hemorrhage. Other signs include loss of appetite, slow growth, loss of weight, skin thickening, suppressed keratinization, increased blood clotting time, reduced erythrocyte count, enteritis, congenital abnormalities, and conjunctivitis. Degenerative atrophy, fatty infiltration, and reduced function of liver and kidneys are typical. Endocrine effects related to the pituitary, thyroid, pancreas, and ovary have been reported in laboratory animals. Because the availability of natural dietary sources of vitamin A and its precursor carotenoids is seasonal, periods of dietary excess accompanied by accumulation of body stores are critical to the health and survival of most animals under natural conditions of feeding.

Normal vitamin A metabolism provides protection from toxicity. The conversion of diverse sources of dietary vitamin A activity to the more stable and less toxic ester form (usually retinyl palmitate) is one such means of protection. Transport to the liver in a lipoprotein complex continues this protection. The tremendous storage capacity of the liver affords great protection against toxicity as well as dietary deficiency. It is not uncommon to find concentrations of 500 to 1,000 IU/g in the livers of most species (Kirk, 1962); however, 13,000 to 18,000 IU/g is common in fish livers (Moore, 1957) and 20,000 IU/g has been observed in human livers (Weber et al., 1982). Storage in the ester form affords protection to liver tissue. Controlled release of the alcohol from the liver by hydrolysis of the ester and subsequent complexing with retinol-binding proteins continue to control reactivity and protect against toxicity. Vitamin A toxicity may be viewed as resulting from intakes that overwhelm one or more of these steps or from malfunctions in this protective system in the presence of intakes that are high but would not normally be toxic. Toxic responses are likely when tissues are exposed to free retinol not bound to retinol-binding protein (Smith and Goodman, 1976). In addition to its role as a storage organ, the liver is the site of glucuronide formation, which facilitates the biliary excretion of vitamin A. The liver is also active in the synthesis of the major vitamin A transport protein, retinol-binding protein.

Although the pathways described above have been studied intensively in only a limited number of species, the available data suggest that most, if not all, species share these routes of metabolism. The apparently greater tolerance for vitamin A of ruminants (see Table 3) than for nonruminant animals is supported by the well-documented ability of ruminal microorganisms to destroy large quantities of vitamin A (Mitchell, 1967).

TABLE 2 Research Findings of High Levels of Vitamin A in Animals

Species and No. of Animals	Age or Weight	Administration Amount	Form	Duration	Route	Effect	Reference
Cats, 31	Weanling	3,500,000 IU/kg diet	Retinyl palmitate	10 mon	Gavage	Loss of appetite; lethargy; irritability; exophthalmus; cervical spondylosis	Seawright et al., 1967
Cats, 3	Weanling	15,000,000 IU/kg diet	Retinyl palmitate	29 wk	Oral	Proliferative gingivitis; incisor exfoliation; thin mandibles	Seawright and Hrdlicka, 1974
Cattle, 23	10–21 wk	2,000,000 IU/kg diet	Retinol	12 wk	Oral	Reduced mucopolysaccharides and RNA	Gorgaez et al., 1971
Cattle, 4	3–6 mo	8,800 IU/kg diet	Retinyl palmitate	12 wk	Oral	Reduced intake	Grey et al., 1965
Cattle, 5	3–6 mo	26,400 IU/kg diet	Retinyl palmitate	12 wk	Oral	Osteoporosis	Grey et al., 1965
Cattle, 3	3–5 mo	35,200 IU/kg diet	Retinyl palmitate	8 wk	Oral	Osteoporosis	Grey et al., 1965
Cattle, 16	332 kg	Up to 256,000 IU/kg diet	Retinyl palmitate	168 d	Oral	No toxicity observed	Hale et al., 1962
Cattle, 6	21 d	1,640,000 IU/kg diet	Retinyl acetate	124 d	Oral	Reduced spinal fluid pressure	Hurt el al., 1967
Cattle, 40	Yearling	8,400 IU/kg diet	Retinyl palmitate	280 d	Oral	No toxicity observed	Frey et al., 1947
Cattle	Mature	120,000 IU/kg diet	Retinyl palmitate	90 d	Oral	No toxicity observed	Walker et al., 1949
Cattle, 4	2½ mo	1,750,000 IU/kg diet	Retinyl palmitate	12 wk	Oral	Decreased spinal fluid pressure; elevated heart rate	Hazzard, 1963
Cattle, 4	2½ mo	3,500,000 IU/kg diet	Retinyl palmitate	12 wk	Oral	Decreased spinal fluid pressure; elevated heart rate	Hazzard, 1963
Birds							
Chickens, 24	1–21 d	21,000 IU/kg diet	Retinyl acetate	21 d	Oral	No effect	Pudelkiewicz et al., 1964
Chickens, 24	1–21 d	1,120,000 IU/kg diet	Retinyl acetate	21 d	Oral	Reduced feed	Pudelkiewicz et al., 1964
Chickens, 18	1 d	3,250,000 IU/kg diet	Retinyl palmitate	32 d	Diet	Depressed growth; 78% mortality	McCuaig and Motzok, 1970
Chickens, hens, 18	28 wk	1,000,000 IU/kg diet	Retinyl palmitate	8 wk	Diet	Reduced hematocrit and plasma vitamin E; increased plasma glutathione peroxidase activity; no effects on egg production or survival	Combs, 1976
Chickens, 30	1 d	1,000,000 IU/kg diet	Retinyl palmitate	2 wk	Diet	Reduced growth and vitamin E absorption; increased plasma glutathione peroxidase activity; no effect on survival	Combs, 1976
Chickens, 62	1 d	650,000 IU/kg diet	Retinyl palmitate	24 d	Diet	Increased erythrocyte glutathione peroxidase activity; decreased erythrocyte superoxide dismutase activity; increased plasma clearance and reduced plasma levels of vitamin E	Sklan and Donoghue, 1982

Animal	Age	Dosage	Compound	Duration	Route	Effects	Reference
Chickens, 180	1 d	1,500,000 IU/kg diet	Retinyl acetate	3 wk	Diet	150,000 IU/kg, depressed growth; 1,500,000 IU/kg, reduced bone ash, increased bone	Veltmann et al., 1986
Chickens	1 d	Up to 2,600,000 IU/kg diet	Retinol	5 wk	Diet	Highest-level depressed growth, hematocrit, and plasma Ca; increased plasma activities of acid phosphatase, β-glucuronidase, and arylsulfatase	Taylor et al., 1968
Chickens, 50	1 d	12,000 IU/kg diet	Retinyl acetate	7 wk	Diet	Reduced growth	Jensen et al., 1983
Chickens, 56	1 d	48,000 IU/kg diet	Retinyl acetate	7 wk	Diet	Reduced growth and pigmentation; increased incidence of leg weakness	Jensen et al., 1983
Chickens, 15	1 d	667,000 IU/kg diet	Retinol	4 wk	Oral	Reduced growth; impaired bone development (inhibited osteoblastic activity, thickened epiphyseal plates)	Baker et al., 1967
Chickens, 2	6–7 d	66,000–99,000 IU/kg diet	Cod liver oil	31 d	Oral	Weight loss; skin lesions; skeletal changes	Wolbach and Hegsted, 1952
Chickens, 360	2–8 wk	52,800 IU/kg diet	Carotene, cod liver oil, retinyl acetate	6 wk	Oral	Slight growth retardation	Castano et al., 1951
Ducks, 6	3–22 d	20,000–120,000 IU/kg diet	Retinyl acetate	14 d	Oral	Reduced intake; osteoporosis	Wolbach and Hegsted, 1953
Turkeys, 400	1–28 d	4,000 and 16,000 IU/kg diet	Retinyl palmitate	4 wk	Oral	No effect	Stevens et al., 1983
Turkeys, 200	1–28 d	44,000 IU/kg diet	Retinyl palmitate	4 wk	Oral	Rickets; reduced feed intake	Stevens et al., 1983
Turkeys, 72	1–56 d	70,000 IU/kg diet	Carotene, cod liver oil, retinyl acetate	8 wk	Oral	No gross symptoms	Garcay et al., 1950
Turkeys, 136	1–126 d	16,000 IU/kg diet	Retinyl palmitate	18 wk	Oral	Reduced growth and femur development	Dorr and Balloun, 1976
Dogs, 5	2 mo	75,000 IU/kg diet	Retinyl acetyl	67 d	Oral	Bone changes	Maddock et al., 1949
Goats, 24	Mature, 58–61 kg	186,000 IU/kg diet	Retinyl palmitate	16 wk	Oral	Decreased spinal fluid pressure	Frier et al., 1974
Rabbits, 10	Young, 1 kg	400 IU/g BW	Retinyl palmitate	5 d	Gavage	Depleted cartilage matrix; hair loss	Thomas et al., 1960
Swine, 7	4–8 kg	Up to 1,100,000 IU/kg diet	Retinyl palmitate	8 wk	Oral	220,000 IU/kg, no effect; 440,000, hemorrhage, slow growth, bone damage	Anderson et al. 1966
Swine, 6	Weanling	660,000 IU/kg diet	Retinyl acetate	2 wk	Oral	Reduced spinal fluid pressure	Hart et al., 1966

TABLE 3   Required and Presumed Upper Safe Levels of Vitamin A (IU/kg diet)

Species	Requirement[a]	Presumed Safe Level[b]
Birds		
Chickens, growing	1,500	15,000
Chickens, laying	4,000	40,000
Ducks	4,000	40,000
Geese	1,500–4,000	15,000
Quail	5,000	25,000
Turkeys, growing	4,000	15,000
Turkeys, breeding	4,000	24,000
Cats	10,000	100,000
Cattle, feedlot	2,200	66,000
Cattle, pregnant, lactating or bulls	2,800–3,900	66,000
Dogs	3,333	33,330
Fish		
Catfish	3,333–6,667	33,330
Salmon	2,500	25,000
Trout	2,500–5,000	25,000
Goats	1,500	45,000
Horses	1,600–3,400	16,000
Monkeys	10,000	100,000
Rabbits	580–1,160	16,000
Sheep	940–3,000	45,000
Swine, growing	2,000	20,000
Swine, breeding	4,000	40,000

[a] From the National Research Council (1977, 1978a, 1978b, 1978c, 1978d, 1979, 1981a, 1981b, 1983, 1984a, 1984b, 1985a, 1985b).
[b] For chronic dietary administration.

Data on interaction of other dietary components with potentially toxic intakes of vitamin A are limited. Vitamin A may be viewed as competing with other fat-soluble vitamins at the sites of absorption. At normal intakes of these vitamins, excess vitamin A may cause them to become deficient as components of the toxicity syndrome. Consequently, elevated intakes of vitamins D, E, and K may reduce vitamin A toxicity by restoring their respective adequacies or by interfering with vitamin A assimilation, or both (Vedder and Rosenberg, 1938; McCuaig and Motzok, 1970; Combs, 1976; Sklan and Donoghue, 1982; Stevens et al., 1983). Protein status can have a major influence through the response of the retinol-binding protein systems (Weber et al., 1982). Protein malnutrition reduces circulating retinol-binding protein. Lack of retinol-binding protein may slow removal of vitamin A from the liver and prevent elevation of blood vitamin A in the presence of potentially toxic vitamin A stores (Weber et al., 1982; Ong, 1985).

Concentrations in Tissues

In most species, more than 90 percent of the vitamin A in the body is stored in the liver (Kirk, 1962). Most of the remaining stores are found in the kidneys, fat depots, adrenals, lungs, and blood. Blood contains levels between 20 and 100 μg of vitamin A/100 ml. In normal ranges, blood levels are poorly correlated with either intake or liver stores. Upon depletion of liver stores of vitamin A, blood concentrations will drop sharply to levels between 5 and 20 μg/100 ml. Persistence of concentrations above 100 μg/100 ml is indicative of toxicity (Eaton, 1969).

PRESUMED UPPER SAFE LEVELS

Experiments have not been conducted with appropriate designs for determining the maximal amounts of vitamin A that can be administered without adverse effects. Consequently, the presumed upper safe levels for orally administered vitamin A are necessarily estimates. The presumed upper safe levels summarized in Table 3 represent levels between the minimal requirements recommended by the National Research Council (NRC) and those reported to be toxic in the referenced scientific publications. When administered for long periods, these levels would be expected to substantially increase stores in the liver. However, the levels have not been reported to produce saturation of the storage capacity of the liver, result in above-normal increases in vitamin A blood concentrations, or elevate retinyl esters in the blood above 50 percent.

The levels selected for nonruminants are consistent with recommendations for humans (Nutrition Foundation, 1982; Olson, 1984). They also agree with Canadian Feed Regulations for the most part (Blair, 1985). The levels are consistent with the ability of mammals to increase concentrations of vitamin A activity in colostrum several times more than concentrations normally found in milk (Walker et al., 1949; Branstetter et al., 1973; Mitchell et al., 1975; Tomlinson et al., 1974, 1976). In view of the lack of reported toxicity in most functioning ruminants, the higher safe levels proposed for them are considered conservative. These levels agree well with the Agricultural Research Council (1980) recommendations for ruminants. The values allow for a wide safety factor in providing requirements or stimulating accumulation of stores.

The biochemical mechanism for vitamin A toxicity is not known. Efforts to assign toxicity to a portion of the retinol molecule have also been unsuccessful.

Absorption of intact carotene is genetically controlled among species. For example, yellow fat species such as cattle and poultry absorb more carotene than white fat species such as sheep and swine. Absorption is also genetically controlled within species. Jersey and Guernsey cows, for instance, put much more carotene in milk than Holsteins. The varying conversions of carotene to vitamin A by the intestine and perhaps other organs cause these differences. High intake of carotenoids from natural feedstuffs does not produce vitamin A toxicity. Carotenosis is not a practical problem in domestic animals. It produces yellowing of the skin but few other adverse signs in humans. In poultry, carotenosis is useful in producing desired color in egg yolks.

SUMMARY

1. Vitamin A is required for normal vision, growth, reproduction, and epithelial tissues in all vertebrates.

2. Excess vitamin A has been demonstrated to have toxic effects in most species studied. However, the excess administered has usually been 10 to 1,000 times the dietary requirements.

3. Presumed upper safe levels are 4 to 10 times the nutritional requirements for nonruminant animals, including birds and fishes, and about 30 times the nutritional requirements for ruminants.

REFERENCES

Agricultural Research Council. 1980. The Nutrient Requirements of Ruminant Livestock. Farnham Royal, England: Commonwealth Agricultural Bureaux.

Agricultural Research Council. 1981. The Nutrient Requirements of Pigs. Farnham Royal, England: Commonwealth Agricultural Bureaux.

Anderson, M. D., V. C. Speer, J. T. McCall, and V. W. Hays. 1966. Hypervitaminosis A in the young pig. J. Anim. Sci. 25:1123.

Baker, J. R., J. M. Howell, and J. N. Thompson. 1967. Hypervitaminosis A in the chick. Br. J. Exp. Pathol. 48:407.

Bauernfeind, J.C. 1980. The Safe Uses of Vitamin A. Washington, D.C.: International Vitamin A Consultative Group. 44 pp.

Bauernfeind, J. C., H. Newmark, and M. Brin. 1974. Vitamin A and E nutrition via intramuscular or oral route. Am. J. Clin. Nutr. 27:234.

Blair, R. 1985. Canadian vitamin ranges for poultry, swine examined. Feedstuffs 57(19):73.

Branstetter, R. F., R. E. Tucker, G. E. Mitchell, Jr., J. A. Boling, and N. W. Bradley. 1973. Vitamin A transfer from cows to calves. Int. J. Vit. Nutr. Res. 43:142.

Castano, F. F., R. V. Boucher, and E. W. Callenbach. 1951. Utilization by the chick of vitamin A from different sources. J. Nutr. 45:131.

Combs, G. F., Jr. 1976. Differential effects of high dietary levels of vitamin A on the vitamin E-selenium nutrition of young and adult chickens. J. Nutr. 106:967.

Dorr, P., and S. L. Balloun. 1976. Effect of dietary vitamin A, ascorbic acid and their interaction on turkey bone mineralization. Br. Poult. Sci. 17:581.

Eaton, H. D. 1969. Chronic bovine hypo- and hypervitaminosis A and cerebrospinal fluid pressure. Am. J. Clin. Nutr. 22:1070–1080.

Frier, H. I., E. J. Gorgaez, R. C. Hall, Jr., A. M. Gallina, J. E. Rousseau, H. D. Eaton, and S. W. Nielsen. 1974. Formation and absorption of cerebrospinal fluid in adult goats with hypo- and hypervitaminosis A. Am. J. Vet. Res. 35:45.

Frey, P. R., R. Jensen, and A. E. Connell. 1947. Vitamin A intake in cattle in relation to hepatic stores and blood levels. J. Nutr. 34:421.

Goodman, D. S. 1980. Vitamin A metabolism. Fed. Proc. 39:2716.

Gorgaez, E. J., J. E. Rousseau, Jr., H. I. Frier, R. C. Hall, Jr., and H. D. Eaton. 1971. Composition of the dura mater in chronic bovine hypervitaminosis A. J. Nutr. 101:1541.

Grey, R. M., S. W. Nielsen, J. E. Rousseau, Jr., M. C. Calhoun, and H. D. Eaton. 1965. Pathology of skull, radius and rib of hypervitaminosis A of young calves. Pathol. Vet. 2:446.

Gurcay, R., R. V. Boucher, and E. W. Callenbach. 1950. Utilization of vitamin A by turkey poults. J. Nutr. 41:565.

Hale, W. H., F. Hubbert, Jr., R. E. Taylor, T. A. Anderson, and B. Taylor. 1962. Performance and tissue vitamin A levels in steers fed high levels of vitamin A. Am. J. Vet. Res. 23:992.

Hayes, K. C., and D. M. Hegsted. 1973. Toxicity of the vitamins. Pp. 235–253 in Toxicants Occurring Naturally in Foods. Washington, D.C.: National Academy of Sciences.

Hazzard, D. G. 1963. Chronic hypervitaminosis A in the bovine. Ph.D. dissertation. University of Connecticut.

Hurt, H. D., R. C. Hall, Jr., M. C. Calhoun, J. E. Rousseau, Jr., H. D. Eaton, R. E. Wolke, and J. J. Lucas. 1966. Chronic hypervitaminosis A in weanling pigs. J. Anim. Sci. 25:891.

Hurt, H. D., H. D. Eaton, J. E. Rousseau, Jr., and R. C. Hall, Jr. 1967. Rates of formation and absorption of cerebrospinal fluid in chronic hypervitaminosis A. J. Dairy Sci. 50:1941.

Jensen, L. S., D. L. Fletcher, M. S. Lilburn, and Y. Akiba. 1983. Growth depression in broiler chicks fed high vitamin A levels. Nutr. Rep. Int. 28:171.

Kirk, J. E. 1962. Variations with age in the tissue content of vitamins and hormones. Vit. Horm. 20:67.

Maddock, S. L., S. Wolbach, and S. Maddock. 1949. Hypervitaminosis A in the dog. J. Nutr. 39:117.

McCuaig, L. W., and I. Motzok. 1970. Excessive dietary vitamin E: Alleviation of hypervitaminosis A and lack of toxicity. Poult. Sci. 49:1050.

Mitchell, G. E., Jr. 1967. Vitamin A nutrition of ruminants. J. Am. Vet. Med. Assoc. 151:430.

Mitchell, G. E., Jr., P. V. Rattray, and J. B. Hutton. 1975. Vitamin A alcohol and vitamin A palmitate transfer from ewes to lambs. Int. J. Vit. Nutr. Res. 45:299.

Moore, T. 1957. Vitamin A. Amsterdam: Elsevier.

National Research Council. 1977. Nutrient Requirements of Rabbits. 2nd rev. ed. Washington, D.C.: National Academy Press.

National Research Council. 1978a. Nutrient Requirements of Cats. Rev. ed. Washington, D.C.: National Academy Press.

National Research Council. 1978b. Nutrient Requirements of Dairy Cattle. 5th rev. ed. Washington, D.C.: National Academy Press.

National Research Council. 1978c. Nutrient Requirements of Horses. 4th rev. ed. Washington, D.C.: National Academy Press.

National Research Council. 1978d. Nutrient Requirements of Nonhuman Primates. Washington, D.C.: National Academy Press.

National Research Council. 1979. Nutrient Requirements of Swine. 8th rev. ed. Washington, D.C.: National Academy Press.

National Research Council. 1981a. Nutrient Requirements of Coldwater Fishes. Washington, D.C.: National Academy Press.

National Research Council. 1981b. Nutrient Requirements of Goats: Angora, Dairy, and Meat Goats in Temperate and Tropical Countries. Washington, D.C.: National Academy Press.

National Research Council. 1983. Nutrient Requirements of Warmwater Fishes and Shellfishes. Rev. ed. Washington, D.C.: National Academy Press.

National Research Council. 1984a. Nutrient Requirements of Beef Cattle. 6th rev. ed. Washington, D.C.: National Academy Press.

National Research Council. 1984b. Nutrient Requirements of Poultry. 8th rev. ed. Washington, D.C.: National Academy Press.

National Research Council. 1985a. Nutrient Requirements of Dogs. Rev. ed. Washington, D.C.: National Academy Press.

National Research Council. 1985b. Nutrient Requirements of Sheep. 6th rev. ed. Washington, D.C.: National Academy Press.

Nieman, C., and H. J. Klein Obbink. 1954. The biochemistry and pathology of hypervitaminosis A. Vit. Horm. (N.Y.) 12:69.

Nutrition Foundation. 1982. The pathophysiological basis of vitamin A toxicity. Nutr. Rev. 40:272.

Olson, J. A. 1969. Metabolism and function of vitamin A. Fed. Proc. 28:1670.

Olson, J. A. 1984. Vitamin A. Pp. 176–191 in Present Knowledge in Nutrition. Washington, D.C.: The Nutrition Foundation, Inc.

Olson, J. A., and M. R. Lakshmanan. 1969. Enzymatic transformations of vitamin A, with particular emphasis on carotenoid cleavage. Pp. 213–226 in The Fat Soluble Vitamins. Madison, Wis.: University of Wisconsin Press.

Ong, D. E. 1985. Vitamin A binding proteins. Nutr. Rev. 43:225.

Ong, D. E., and F. Chytil. 1983. Vitamin A and cancer. Vit. Horm. 40:105.

Pitt, G. A. J. 1985. Vitamin A. Pp. 1–75 in Fat-Soluble Vitamins, A. T. Diplock, ed. Lancaster, Pa.: Technomic Publishing.

Pudelkiewicz, W. J., L. Webster, G. Olson, and L. D. Matterson. 1964. Some physiological effects of feeding high levels of vitamin A acetate to chicks. Poult. Sci. 45:1157.

Seawright, A. A., and J. Hrdlicka. 1974. Pathogenetic factors in tooth loss in young cats on a high daily oral intake of vitamin A. Aust. Vet. J. 50:133.

Seawright, A. A., P. B. English, and R. J. W. Gartner. 1967. Hypervitaminosis A and deforming cervical spondylosis of the cat. J. Comp. Pathol. 77:29.

Sklan, D., and S. Donoghue. 1982. Vitamin E response to high dietary vitamin A in the chick. J. Nutr. 112:759.

Smith, F. R., and D. W. Goodman. 1976. Vitamin A transport in human vitamin A toxicity. N. Engl. J. Med. 294:805.

Stevens, V. I., R. Blair, and C. Riddell. 1983. Dietary levels of fat, calcium and vitamins A and D_3 as contributory factors to rickets in poults. Poult. Sci. 62:2073.

Taylor, T. G., M. L. Morris, and J. Kirkley. 1968. Effects of dietary excesses of vitamins A and D on some constituents of the blood of chicks. Br. J. Nutr. 22:713.

Thomas, L., R. T. McCluskey, J. L. Potter, and G. Weissman. 1960. Comparison of the effects of papain and vitamin A on cartilage. I. The effect in rabbits. J. Exp. Med. 111:705.

Tomlinson, J. E., G. E. Mitchell, Jr., N. W. Bradley, R. E. Tucker, J. A. Boling, and G. T. Schelling. 1974. Transfer of vitamin A from bovine liver to milk. J. Anim. Sci. 39:813.

Tomlinson, J. E., R. W. Hemken, G. E. Mitchell, Jr., and R. E. Tucker. 1976. Mammary transfer of vitamin A alcohol and ester in lactating dairy cows. J. Dairy Sci. 59:607.

Ullrey, D. E. 1972. Biological availability of fat soluble vitamins: Vitamin A and carotene. J. Anim. Sci. 35:648.

Vedder, E. B., and C. Rosenberg. 1938. Concerning the toxicity of vitamin A. J. Nutr. 16:57.

Veltmann, J. R., Jr., L. S. Jensen, and G. N. Rowland. 1986. Excess dietary vitamin A in the growing chick: Effect of fat source and vitamin D. Poult. Sci. 65:153.

Wald, G. 1968. Molecular basis of visual excitation. Science 162:230.

Walker, D. M., S. Y. Thompson, S. Bartlett, and S. K. Kon. 1949. The effect of diet during pregnancy on the vitamin A and carotene content of colostrum of cows and heifers and on the reserves of the calf. Rep. 12th Int. Dairy Congr., Stockholm, Sweden 1:83–88.

Weber, F. L., Jr., G. E. Mitchell, Jr., D. E. Powell, B. J. Reiser, and J. G. Banwell. 1982. Reversible hepatotoxicity associated with hepatic vitamin A accumulation in a protein-deficient patient. Gastroenterology 82:118–123.

Wolbach, S. B., and D. M. Hegsted. 1952. Hypervitaminosis A and the skeleton of growing chicks. Am. Med. Assoc. Arch. Pathol. 54:30.

Wolbach, S. B., and D. M. Hegsted. 1953. Hypervitaminosis A in young ducks. Am. Med. Assoc. Arch. Pathol. 55:47.

Vitamin D

Vitamin D has been known since 1920 as a chemical and nutritional entity. Not until after the late 1960s, however, has the biochemical basis of its physiological role been at least partially defined. Since then, research in vitamin D metabolism has resulted in new and important information regarding its role in calcium and phosphorus metabolism. Research has also given insights into the value of vitamin D in clinical medicine related to abnormal mineral metabolism, endocrinology, and nutrition. This information has been fostered by the development of high specific activity, radiolabeled vitamin D, and more precise isolation and identification methods to study vitamin D metabolism in greater detail.

NUTRITIONAL ROLE

Vitamin D can be considered a vitamin only in the sense that, under modern farming conditions, many animals are raised in total confinement with little or no exposure to natural sunlight. Adequate sunlight results in the production of sufficient vitamin D_3 from 7-dehydrocholesterol in the skin. Hence, vitamin D_3 is not required in the diet if sufficient amounts of sunlight are received.

Lack of adequate photoproduction of vitamin D_3 or inadequate dietary supplementation of vitamin D leads to the failure of bones to calcify normally. This metabolic disease is known as rickets in the young and osteomalacia in adults. Once this deficiency was recognized, the dietary supplementation of vitamin D became a commonly accepted nutritional practice. The dietary requirements for most animal species are in the range of 200 to 1,200 IU/kg of diet.

FORMS OF THE VITAMIN

The vitamin D sterols that are used in human and veterinary medicine and their relative biologic poten-cies in mammals are listed in Table 4. Toxicity has been reported with many of these. The most common occurrences result from the use of vitamin D_2 or vitamin D_3 (Figure 2). Vitamin D toxicity has also occurred with ingestion of certain plants containing a water-soluble glycoside of 1,25-$(OH)_2$-D_3 (Hughes et al., 1977b).

ABSORPTION AND METABOLISM

Many excellent reviews have been written on the metabolism and function of vitamin D (Haussler and Mc-Cain, 1977; Norman, 1979; Stern, 1980; Norman et al., 1982; Horst and Reinhardt, 1983; DeLuca, 1984; Horst, 1986) and should be consulted for in-depth information. The following discussion, therefore, is limited to some of the key events leading to the in vivo activation of vitamin D.

Because it is fat soluble, vitamin D is absorbed with other neutral lipids via chylomicrons into the lymphatic system of mammals or the portal circulation of birds and fishes. The two major natural sources of vitamin D are cholecalciferol (vitamin D_3, which occurs in animals) or ergocalciferol (vitamin D_2, which occurs predominantly in plants). Vitamin D (absence of a subscript implies either vitamin D_2 or vitamin D_3) either ingested or produced in the skin is carried through the circulatory system to the liver, where it is converted to 25-hydroxyvitamin D (25-OH-D). This metabolite is the major circulating form under normal conditions and during vitamin D excess (Horst and Littledike, 1982; Littledike and Horst, 1982). For some time, it was considered to be the metabolically active form of vitamin D (DeLuca, 1971). It is now known to be the precursor to 1,25-dihydroxyvitamin D (1,25-$(OH)_2$-D), the active metabolite that is produced almost exclusively in the kidney. This metabolite functions with the parathyroid hor-

11

TABLE 4   Vitamin D Sterols Used in Human and Animal Nutrition: Their Relative Antirachitic Potencies and Duration of Effects Following Withdrawal in Mammals

Name	Synonym	Relative Potency[a]	Duration (weeks)	Comments
Vitamin D	Calciferol	1	6–18	Either vitamin D_2 or vitamin D_3
Vitamin D_3	Cholecalciferol	1	6–18	Animal form: produced by irradiation of 7-dehydrocholesterol
Vitamin D_2	Ergocalciferol	1	6–18	Plant form: produced by irradiation of ergosterol
Dihydrotachy-sterol	DHT	0.05–0.1	1–3	Sterol generated during irradiation of ergosterol
25-OH-D_3	Calcidiol	2–5	4–12	Liver metabolite of vitamin D_3
1,25-$(OH)_2$-D_3	Calcitriol	5–10	0.2–0.8	Kidney metabolite of 25-hydroxy-vitamin D_3
lα-OH-D_3	α-Calcidiol	5–10	0.3–1.0	Synthetic analogue

[a]Adapted from Parfitt (1980).

mone (PTH) to bring about blood calcium and phosphorus homeostasis. The PTH acts with 1,25-$(OH)_2$-D to regulate plasma calcium and phosphorus concentrations. The hormone is also an important mediator of the renal production of 1,25-$(OH)_2$-D (that is, of the 25-OH-D-lα-hydroxylase). Other factors, however, can influence the biosynthesis of 1,25-$(OH)_2$-D, some of which are listed in Figure 3. Once formed, 1,25-$(OH)_2$-D binds to a specific receptor in the enterocyte nucleus and initiates events leading to a stimulation in calcium and phosphorus absorption (Norman et al., 1982). Also, 1,25-$(OH)_2$-D acting with PTH mediates the resorption of bone with the release of calcium and phosphorus (DeLuca, 1984). A series of new discoveries has made it apparent that 1,25-$(OH)_2$-D plays a much wider role in biology than was thought. A variety of tissues not regarded to participate in mineral and skeletal homeostasis have been found to possess specific receptors for 1,25-$(OH)_2$-D (Norman et al., 1982).

There are many metabolites of vitamin D that circulate in plasma other than 25-OH-D and 1,25-$(OH)_2$-D. Table 4 lists some of the vitamin D_3 and vitamin D_3 metabolites. One metabolite, 24,25-dihydroxyvitamin D (24,25-$(OH)_2$-D), has also been considered as a biologically active vitamin D metabolite. Although the physiologic significance of 24,25-$(OH)_2$-D is not yet understood, it has been proposed to have a role in the formation of bone (Norman, 1980).

FIGURE 2   Chemical structures of vitamin D_3 and vitamin D_2.

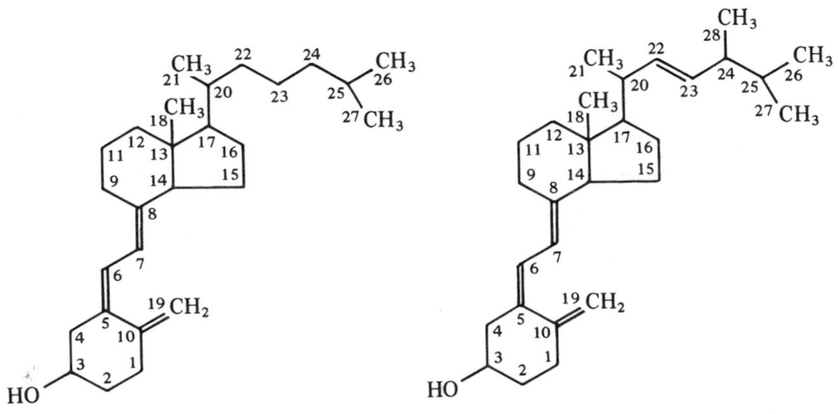

Vitamin D_3
(cholecalciferol)

Vitamin D_2
(ergocalciferol)

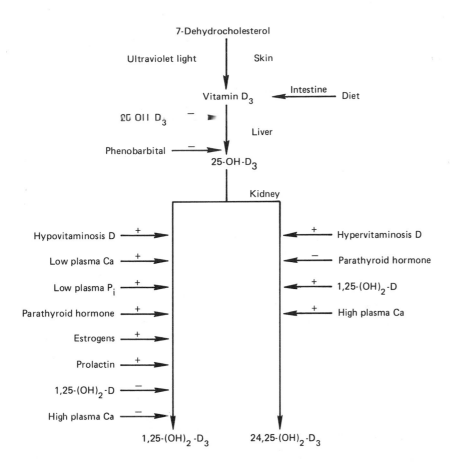

FIGURE 3 Factors regulating 1,25-dihydroxyvitamin D and 24,25-dihydroxyvitamin D biosynthesis.

HYPERVITAMINOSIS

Putscher noted vitamin D toxicity as early as 1929. Toxicity has been described in many species, including humans. Accidental toxicity has been reported in various species of animals, including monkeys, dogs, cattle, horses, pigs, and chinchillas.

Although its toxicity in humans has been known for more than 40 years, the vitamin's significance in veterinary medicine has drawn greater attention in connection with massive administration to prevent milk fever in dairy cows.

Clinical Signs

Many investigators have described the clinical signs of hypervitaminosis D in mammals. Cole et al. (1957) reported that cows receiving 30 million IU of vitamin D_2 orally for 11 days developed anorexia, reduced rumination, depression, premature ventricular systoles, and bradycardia. Kent et al. (1958) observed in monkeys weight loss, anorexia, elevated blood urea nitrogen (BUN), diarrhea, anemia, and upper respiratory infections. In pigs, Chineme et al. (1976) described anorexia, stiffness, lameness, arching of the back, polyuria, and aphonia.

It is generally assumed that vitamin D_2 and vitamin D_3 are equally potent in most mammals. In certain animals, however, it is quite clear that there are substantial differences between the two sterols. In birds (Chen and Bosmann, 1964) and in New World primates (Hunt et al., 1967), vitamin D_3 is substantially more active than vitamin D_2. It has generally been assumed that vitamins D_2 and D_3 are equally active in Old World monkeys in augmenting calcium absorption and preventing osteomalacia. However, when large and potentially toxic doses were administered orally to rhesus monkeys (Hunt et al., 1972), vitamin D_3 was more toxic. Hypercalcemia, extensive soft tissue calcification, and death occurred in many animals. By contrast, the administration of vitamin D_2 produced hypercalcemia to a lesser degree. Animals survived, and soft tissue calcification was absent or only mild. Similarly, Harrington and Page (1983) found vitamin D_3 more hypercalcemic and overtly toxic to horses than vitamin D_2.

The development of methods to measure vitamin D and its metabolites in plasma (Horst et al., 1981) has provided insight into the possible mechanism of vitamin D toxicity and also has provided information regarding the metabolic bases of the differences in toxicity between vitamins D_2 and D_3.

As stated earlier, the predominant vitamin D form in

plasma following vitamin D overdose is 25-OH-D. This metabolite circulates normally at 30 to 50 ng/ml in most species (Horst and Littledike, 1982). However, during vitamin D intoxication, it increases from 200 to 400 ng/ml (Littledike and Horst, 1982). When circulating at very high concentrations, 25-OH-D can compete effectively with 1,25-(OH)$_2$-D for receptors in the intestine and bone. Therefore, during vitamin D toxicosis, 25-OH-D can induce actions usually attributed to 1,25-(OH)$_2$-D. Thus, 25-OH-D is believed to be the critical factor in vitamin D intoxication.

When equal amounts of vitamin D$_3$ and vitamin D$_2$ are presented together in diets of mammals, the predominant circulating form of the vitamin is usually 25-OH-D$_3$ rather than 25-OH-D$_2$ (Horst et al., 1982). Similarly, in toxicity experiments where vitamin D$_2$ was less toxic than vitamin D$_3$, the metabolite 25-OH-D$_2$ was found to be present at lower plasma concentrations than was 25-OH-D$_3$ (Harrington and Page, 1983). Therefore, the difference in toxicity between these vitamins is probably attributable to the less efficient metabolism of vitamin D$_2$ to its more active metabolites, particularly 25-OH-D$_2$. In most species, plasma concentrations of 1,25-(OH)$_2$-D decrease during toxicosis (Hughes et al., 1977a). However, there are differences between species in this response. For example, bovine species show substantial increases in plasma 1,25-(OH)$_2$-D following intramuscular doses of vitamin D$_3$ in massive amounts (15 million IU) (Horst and Littledike, 1979). Therefore, vitamin D toxicity in ruminants may be partially a response to elevated 1,25-(OH)$_2$-D.

A summary of the major pathogenic factors involved in vitamin D toxicity is shown in Figure 4. Treatment with excess vitamin D or 25-OH-D stimulates intestinal absorption of calcium and, to a lesser degree, augments intestinal phosphate transport. Bone resorption of calcium is increased. The overall effect is an increase in serum calcium and reduction in PTH. With modest hypercalcemia, glomerular filtration rate (GFR) may remain stable, and hypercalciuria may be substantial because of the increased filtered load of calcium and the reduction of tubular reabsorption of calcium as a result of reduced PTH secretion. When GFR is maintained, serum calcium may only be modestly elevated by 10 to 20 percent. There is an increased risk of nephrolithiasis because of the hypercalciuria, however. With further increases in serum calcium level, the GFR decreases. This decrease is due to the potentiating action of calcium on angiotensin II-mediated vasoconstriction of renal afferent arterioles. A further, rapid increase in serum calcium might then occur due to the decrease in filtered calcium and the subsequent fall in urinary calcium. Polyuria, along with vomiting (in nonruminants), may cause the extracellular fluid volume to be reduced, which would further contribute to reduced renal function. Thus, reduced renal function is the major event

FIGURE 4 Scheme for the pathogenesis of vitamin toxicosis. The abbreviations and their meanings are: Ca, calcium; GFR, glomerular filtration rate; 25-OH-D, 25-hydroxyvitamin D; P, phosphate; PTH, parathyroid hormone; T$_m$, tubular maximum.

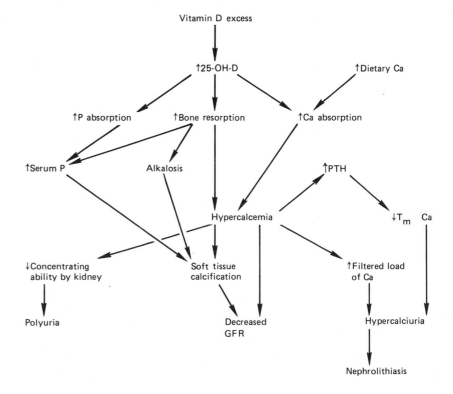

that leads to the total loss of control of calcium homeostasis and to the severity of hypercalcemia during vitamin D intoxication.

Soft Tissue Changes

Postmortem examination of vitamin D-intoxicated animals generally reveals extensive cardiovascular and kidney mineralization. In cows and sheep given toxic doses of vitamin D_3, the collecting tubules of the medulla are the major mineralization sites. The cortex is a minor mineralization site (Capen et al., 1966; Simesen et al., 1978). However, the cortex and papillae are major mineralization sites in 1α-OH-D_3 (a precursor to 1,25-(OH)$_2$-D_3) toxicity in sheep (Simesen et al., 1978). Cardiovascular lesions are primarily located in the aorta, stomach arteries, aortic valves, aortic arch, large arterial bifurcations, and area around the openings of small vessels.

Mineralization within the respiratory tract is also one of the most frequent lesions. Kent et al. (1958) described calcification of the basement membrane of small bronchi, alveolar ducts, and bronchial cartilage in monkeys. Chineme et al. (1976) described calcification of the alveolar septa, bronchial submucosa, and walls of arterioles in pigs.

Kent et al. (1958) also showed that after kidney lesions, calcification of the salivary glands (calcification occurs twice as often in the submaxillary gland as in the parotid) was the next earliest and most frequent lesion. Chineme et al. (1976) have described calcification in the mucosa and muscularis mucosae of the stomach of the pig's stomach.

Lesions That Are Not Associated with Hypercalcemia

The possibility that long-term treatment with high levels of vitamin D or an active sterol may cause tissue damage, particularly to the kidney, in the absence of hypercalcemia has been the subject of considerable speculation. While the bulk of information suggests that hypercalcemia is the sine qua non to manifestation of vitamin D intoxication, there are studies in experimental animals suggesting that mild ultrastructural abnormalities occur before the appearance of hypercalcemia or the deposition of calcium in tissues (Manston and Payne, 1964). Also, there have been reports of the development of nephrocalcinosis and hypercalciuria in humans without known hypercalcemia treated with vitamin D or dihydrotachysterol (Dinkel, 1966). A retrospective evaluation of 27 patients with hypoparathyroidism treated with pharmacologic doses of vitamin D or dihydrotachysterol has suggested that renal function did decrease in certain patients in the absence of hyper-

calcemia (Parfitt, 1977). In 5 patients, the development of nephrocalcinosis correlated with the frequency and severity of documented hypercalcemia. Nephrocalcinosis developed in 3 other patients, however, in whom there was no correlation with the frequency or severity of hypercalcemia nor tendency toward hyperphosphatemia.

Factors Affecting Toxicity

The severity of the effects and pathogenic lesions in vitamin D intoxication depend upon such factors as the type of vitamin D (vitamin D_2 versus vitamin D_3), the dose, the functional state of the kidneys, and the composition of the diet. Vitamin D toxicity is enhanced by a rich dietary supply of calcium and phosphorus, and is reduced when the diet is low in calcium (Hines et al., 1985). Toxicity is also reduced when the vitamin is accompanied by high intakes of vitamin A or by thyroxin injections (Payne and Manston, 1967). The route of administration also influences toxicity. Parenteral administration of 15 million IU of vitamin D_3 in a single dose caused toxicity and death in many pregnant dairy cows (Littledike and Horst, 1982). On the other hand, oral administration of 20 to 30 million IU of vitamin D_2 daily for 7 days resulted in little or no toxicity in pregnant dairy cows (Hibbs and Pounden, 1955). Napoli et al. (1983) have shown that rumen microbes are capable of metabolizing vitamin D to the inactive 10 keto-19-nor vitamin D. Parenteral administration would circumvent the deactivation of vitamin D by rumen microbes and may partially explain the difference in toxicity between oral and parenteral vitamin D.

Various measures have been used in human medicine for treatment of vitamin D toxicity. These measures are mainly concerned with hypercalcemia management. Vitamin D withdrawal is obviously indicated. It is usually not immediately successful, however, due to the long plasma half-life of vitamin D (5 to 7 days) and 25-OH-D (20 to 30 days). This is in contrast to the short plasma half-life of 1α-OH-D_3 (1 to 2 days) and 1,25-(OH)$_2$-D_3 (4 to 8 hours). Because intestinal absorption of calcium contributes to hypercalcemia, a prompt reduction in dietary calcium is indicated. Sodium phytate, an agent that reduces intestinal calcium absorption, has also been used successfully in vitamin D toxicity management in monogastrics (Recker et al., 1979). This treatment would be of little benefit to ruminants due to the presence of rumen microbial phytases. There have also been reports that calcitonin (West et al., 1971), glucagon (Ulbrych-Jablonska, 1972), and glucocorticoid therapy (Streck et al., 1979) reduce serum calcium levels resulting from vitamin D intoxication.

Concentrations in Milk and Liver

Hollis et al. (1981), Reeve et al. (1983), Kunz et al. (1984), and McDermott et al. (1985) have reported the distribution of vitamin D and vitamin D metabolites in milk and colostrum of normal dairy cows. Estimates have also been made following administration of pharmacologic amounts of vitamin D or vitamin D metabolites to dairy cows. Colostrum from cows receiving normal dietary amounts (10,000 to 50,000 IU/day) of vitamin D had 200 to 300 IU of vitamin D activity/liter. Normal milk had 40 to 50 IU/liter. Colostrum from cows receiving 30 million IU of vitamin D_2 before parturition contained 13,000 IU/liter. Normal milk taken 6 days following parturition contained 2,400 IU/liter (Hibbs and Pounden, 1955). Daily feeding of 162,000 IU of vitamin D_3 from cod liver oil led to an 11-fold increase in vitamin D activity in milk (Krauss and Bethke, 1937).

Hollis et al. (1983) published one of the first reports regarding the concentration of vitamin D and vitamin D metabolites in milk from dairy cows and humans receiving large parenteral or enteral doses of vitamin D. In cows that received 125 mg (15 million IU) of vitamin D_3, vitamin D_3 and 25-OH-D_3 concentrations in plasma increased significantly 20 days before parturition. This increase was reflected by similar increases in colostrum and milk concentrations of these sterols (Figures 5 and 6). Similarly, in mothers given supplementations of

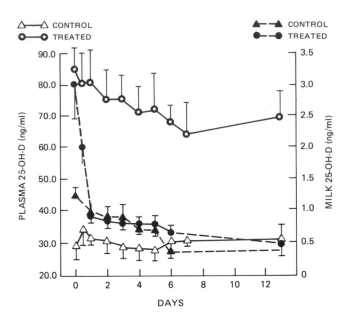

FIGURE 6   Relationship between plasma and milk levels of 25-OH-D in the cow. Treatments were the same as those described in Figure 5 (Hollis et al., 1983).

2,000 IU of vitamin D_2 during late gestation and early lactation, the milk concentrations of vitamin D and 25-OH-D were significantly elevated from those observed in milk from mothers given normal supplementations (400 IU/day). In both cases, the concentrations of 24,25-$(OH)_2$-D and 1,25-$(OH)_2$-D_3 were not elevated. When normal cows were treated with 400 μg of 1,25-$(OH)_2$-D_3 parenterally, 1,25-dihydroxyvitamin D_3 was elevated in milk, however (Hollis et al., 1983).

Vitamin D activity was also elevated in cows' livers following dietary supplementation with 250,000 IU/day for 2 to 3 weeks (Quarterman et al., 1964). At the time of sacrifice, the vitamin D activity had increased to 2,700 IU/100 g of tissue compared to 21 IU/100 g of tissue in the control group. Following withdrawal of the vitamin D, the activity in the liver had decreased to normal levels within 2 to 3 weeks.

PRESUMED UPPER SAFE LEVELS

Existing data for several of the domestic species do not allow precise estimates to be made for maximum vitamin D tolerance levels. Rather, most of the experiments to date reviewed (Table 5) have addressed the clinical consequences of vitamin D toxicosis.

Several factors, such as the chemical form (vitamin D_2 or vitamin D_3), species, dietary intake of calcium and phosphorus, route of administration, and duration of treatment, can influence the maximum tolerable levels

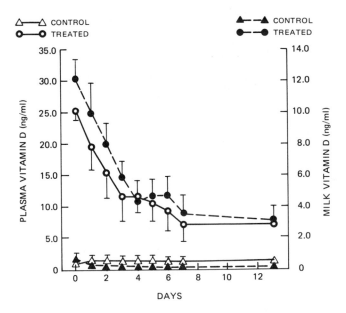

FIGURE 5   Relationship between plasma and milk levels of vitamin D in the cow. Treated animals were injected intramuscularly with 5 million IU of vitamin D_3 approximately 20 days before parturition. Control animals were maintained on a diet containing 4,000 IU of vitamin D_3/day (Hollis et al., 1983).

TABLE 5 Research Findings of High Levels of Vitamin D in Animals

Species and No. of Animals	Age or Weight	Administration Amount[a]	Form	Duration	Route	Effect	Reference
Birds							
Chickens, 10	10 d	300 μg/3 d	Vitamin D$_3$	4 doses	Oral	Nephrocalcinosis	Ratzkowski et al., 1982
Chickens, 10	10 d	300 μg/3 d	Vitamin D$_3$	4 doses	SC	Nephrocalcinosis	Ratzkowski et al., 1982
Chickens laying hens, 10		6.8 μg/kg diet	1α-OH-D$_3$	22 wk	Diet	Anorexia; reduced eggshell quality; reduced egg production	Soares et al., 1982
Chickens, laying hens, 10		6–12 μg/kg diet	25-OH-D$_3$	22 wk	Diet	No effect	Soares et al., 1982
Chickens, 15	1 d	1 mg/kg diet	Vitamin D$_3$	14 d	Diet	No effect	Morrissey et al., 1977
Chickens, 15	1 d	10 mg/kg diet	Vitamin D$_3$	14 d	Diet	Renal tubular calcification; muscular atrophy; emaciation	Morrissey et al., 1977
Chickens, 15	1 d	0.01 mg/kg diet	25-OH-D$_3$	14 d	Diet	No effect	Morrissey et al., 1977
Chickens, 15	1 d	0.1 mg/kg diet	25-OH-D$_3$	14 d	Diet	Renal tubular calcification; muscular atrophy; emaciation	Morrissey et al., 1977
Chickens, 12	1 d	12.5 mg/kg diet	Vitamin D$_3$	4–6 wk	Diet	Hypercalcemia; hypophosphatemia; weight loss	Taylor et al., 1968
Chickens, 12	1 d	12.5 mg/kg diet and 2,600,000 IU/kg diet	Vitamin D$_3$ and retinol	4–6 wk	Diet	Weight loss; normocalcemia; normophosphatemia	Taylor et al., 1968
Japanese quail, 24	6 mo	3 mg/kg diet	Vitamin D$_3$	28 d	Diet	No effect	Stevens and Blair, 1985
Turkeys, 10	1 d	2.25 mg/kg diet	Vitamin D$_3$	21–25 d	Diet	No effect	Metz et al., 1985
Turkeys, 10	1 d	22.5 mg/kg diet	Vitamin D$_3$	21–25 d	Diet	Growth depression; renal tubular calcinosis	Metz et al., 1985
Cattle, 1	6 yr	750 mg/d	Vitamin D$_2$	21 d	Diet	Anorexia; cardiac irregularity; metastatic calcification	Cox et al., 1957
Cattle, 4	Mature	750 mg/d	Vitamin D$_2$	7 d	Diet	Roughing of aorta	Capen et al., 1966
Cattle, 4	Mature	750 mg/d	Vitamin D$_2$	10 d	Diet	Mineralization of aorta and endocardium	Capen et al., 1966
Cattle, 4	Mature	750 mg/d	Vitamin D$_2$	21 d	Diet	Cardiovascular mineralization; mineralization of kidney medulla	Capen et al., 1966
Cattle, 4	Mature	750 mg/d	Vitamin D$_2$	30 d	Diet	Extensive cardiovascular mineralization	Capen et al., 1966
Cattle	Mature	250 mg	Vitamin D$_3$	Single dose	IV	No effect	Seekles et al., 1961
Cattle	Mature	250 mg	Vitamin D$_3$	Single dose	IM	No effect	Seekles et al., 1961
Cattle	Mature	250 mg/8 d	Vitamin D$_3$	2 doses	IV	Shock; cardiac irregularity	Wilson, 1964
Cattle, 7	Mature	250 mg	Vitamin D$_3$	Single dose	IV	No effect	Greig, 1963
Cattle, 7	Mature	500 mg	Vitamin D$_3$	Single dose	IV	Anorexia; depression; cardiac irregularity; death	Greig, 1963
Cattle, 2	Mature	750 mg	Vitamin D$_3$	Single dose	IV	Death within 2 wk	Greig, 1963
Cattle, 2	Mature	1,000 mg	Vitamin D$_3$	Single dose	IV	Death within 2 wk	Greig, 1963
Cattle, 6	Mature	125 mg	Vitamin D$_3$	Single dose	IM	Cardiovascular mineralization	Payne and Manston, 1967
Cattle, 12	Mature	250 mg	Vitamin D$_3$	Single dose	IM	Cardiovascular mineralization	Payne and Manston, 1967
Cattle, 4	Mature	1,000 mg	Vitamin D$_3$	Single dose	IM	Cardiovascular mineralization	Payne and Manston, 1967

18

TABLE 5—Continued

Species and No. of Animals	Age or Weight	Amount[a]	Form	Duration	Route	Effect	Reference
Cattle, 17	Pregnant	375 mg	Vitamin D_3	Single dose	IM	Anorexia; cardiovascular mineralization; death in 30 d	Littledike and Horst, 1982
Cattle, 2	140 kg	2.1 mg/7 d	1α-OH-D_3	4 doses	IM	Anorexia; cardiovascular calcinosis; cardiac irregularity	Mullan et al., 1979
Cattle, 2	Mature	1 mg/d	25-OH-D_3	14 d	Oral	No effect	Olson et al., 1973
Cattle, 2	Mature	4 mg/d	25-OH-D_3	Single dose	IM	No effect	Olson et al., 1973
Cattle, 2	Mature	8 mg/d	25-OH-D_3	Single dose	IM	No effect	Olson et al., 1973
Cattle, 2	Mature	16 mg/d	25-OH-D_3	Single dose	IM	No effect	Olson et al., 1973
Dogs, males, 18	Mature	0.5–1.0 mg/kg BW	Vitamin D_2	1–3 wk	Oral	Hypercalcemia; cardiovascular and nephrocalcinosis; increased blood pressure	Spangler et al., 1979
Fish							
Catfish		0.5 mg/kg diet	Vitamin D_3	28 wk	Oral	No effect	Andrews et al., 1980
Catfish		1.25 mg/kg diet	Vitamin D_3	28 wk	Oral	Decreased weight gain and feed efficiency	Andrews et al., 1980
Rainbow trout		25 mg/kg diet	Vitamin D_3	24 wk	Oral	No effect	Hilton and Ferguson, 1982
Brook trout-fingerling		93 mg/kg diet	Vitamin D_3	40 wk	Oral	Hypercalcemia; decreased weight gain	Poston, 1968
Foxes	2–6 mo	0.125–0.250 mg/kg BW	Vitamin D_3	3 mo	Oral	Anorexia; hypercalcemia; cardiovascular calcinosis; hyaline muscle damage; death in several animals consuming 0.250 mg/kg BW	Helgebostad and Nordstoga, 1978
Horses, 1	270–321 kg	1.18 mg/kg BW	Vitamin D_2	21 d	Diet	Severe cardiovascular calcinosis; hypercalcemia; hyperphosphatemia	Harrington, 1982
Horses, 2		0.082 mg/kg BW	Vitamin D_3	4 mo	Diet	Kidney and cardiovascular calcinosis; death	Hintz et al., 1973
Horses, 1	231 kg	0.825 mg/kg BW	Vitamin D_2	33 d	Diet	Mild cardiovascular calcinosis; weight loss; hypercalcemia; hyperphosphatemia	Harrington and Page, 1983
Horses, 1	304 kg	0.825 mg/kg BW	Vitamin D_3	33 d	Diet	Severe weight loss; hypercalcemia; hyperphosphatemia; severe cardiovascular calcinosis	Harrington and Page, 1983
Horses, 1	270–321 kg	0.232 mg/kg BW	Vitamin D_2	21 d	Diet	No effect	Harrington, 1982
Horses, 1	270–321 kg	0.550 mg/kg BW	Vitamin D_2	21 d	Diet	No effect	Harrington, 1982
Mink, 5	2–6 mo	0.175–0.375 mg/kg BW	Vitamin D_3	2–3 wk	Oral	No effect	Helgebostad and Nordstoga, 1978
Monkeys							
Rhesus monkeys, 2	Young adult	1.25 mg/d	Vitamin D_2	160 d	Oral	Hypercalcemia	Hunt et al., 1972
Rhesus monkeys, 2	Young adult	2.50 mg/d	Vitamin D_2	160 d	Oral	Hypercalcemia	Hunt et al., 1972
Rhesus monkeys, 2	Young adult	5.0 mg/d	Vitamin D_2	160 d	Oral	Hypercalcemia	Hunt et al., 1972

Species, No.	Body weight/age	Compound	Dose	Duration	Route	Effect	Reference
Rhesus monkeys, 2	Young adult	Vitamin D₃	1.25 mg/d	66–160 d	Oral	Hypercalcemia; cardiovascular and soft tissue metabolism; death	Hunt et al., 1972
Rhesus monkeys, 2	Young adult	Vitamin D₃	2.50 mg/d	90 d	Oral	Hypercalcemia; cardiovascular and soft tissue metabolism; death	Hunt et al., 1972
Rhesus monkeys, 2	Young adult	Vitamin D₃	5.0 mg/d	16–123 d	Oral	Hypercalcemia; cardiovascular and soft tissue metabolism; death	Hunt et al., 1972
Squirrel monkeys	600–900 g	Vitamin D₂	1.25 mg/d	42 d	Oral	Slight hypercalcemia	Hunt et al., 1969
Squirrel monkeys	600–900 g	Vitamin D₂	2.50 mg/d	42 d	Oral	Slight hypercalcemia	Hunt et al., 1969
Squirrel monkeys	600–900	Vitamin D₃	1.25 mg/d	20–30 d	Oral	Hypercalcemia; soft tissue mineralization; death	Hunt et al., 1969
Squirrel monkeys	600–900	Vitamin D₃	2.50 mg/d	20–30 d	Oral	Hypercalcemia; soft tissue mineralization; death	Hunt et al., 1969
Rabbits		Vitamin D₃	0.250 mg/kg BW		Diet	Cardiovascular calcinosis	Toda et al., 1983
Sheep, 2	6 mo	Vitamin D₃	1 mg/kg BW, once every 7 d	4 doses	IM	Hypercalcemia; cardiovascular and kidney calcification	Simesen et al., 1978
Sheep, 2	6 mo	1α-OH-D₃	1 μg/kg BW, once every 7 d	4 doses/7 d	IM	Hypercalcemia; cardiovascular and kidney calcification	Simesen et al., 1978
Sheep, 4		Vitamin D₂	12.5 mg/d	6 wk	IM	Cervical scoliosis; kidney calcification	Clegg and Hollands, 1976
Sheep, 15	52 kg	Vitamin D₃	125, 250, and 500 μg/kg diet	16 wk	Oral	Elevated plasma 25-OH-D₃	Smith et al., 1985
Sheep, 10	52 kg	Vitamin D₃	31 and 62 μg/kg diet	16 wk	IM	No effect	Smith et al., 1985
Swine, 7	Weanling	Vitamin D₃	5.5 mg/kg diet	6 wk	Diet	Cardiovascular mineralization	Kamio et al., 1977
Swine, 4	Weanling	Vitamin D₃	0.825 mg/kg diet	8 wk	Diet	No effect	Chineme et al., 1976
Swine, 4	Weanling	Vitamin D₃	4.13 mg/kg diet	8 wk	Diet	Hypercalcemia; hyperphosphatemia; osteoporosis	Chineme et al., 1976
Swine, 4	Weanling	Vitamin D₃	20.6 mg/kg diet	8 wk	Diet	Hypercalcemia; hyperphosphatemia; soft tissue calcinosis	Chineme et al., 1976
Swine, 30	3 wk	Vitamin D₃	7.4 mg/kg diet	5 wk	Diet	Inappetance; decreased rate of growth; rough hair coat; lameness; paralysis	Wren, 1980
Swine, 19	8 wk	Vitamin D₃	6.25 mg/kg diet	4 wk	Diet	Cardiovascular lesions; decreased growth rate	Quarterman et al., 1964
Swine, 35	Fattening hogs and gilts	Vitamin D₃	11.8 mg/kg diet	2 d	Diet	Cardiovascular mineralization; hypercalcemia; death within 2–4 d of some animals	Long, 1984
Swine	2 mo	Vitamin D₃	1.56 mg/kg diet	3 mo	Diet	Atherosclerotic lesions; smooth muscle cell death	Toda et al., 1983

[a] One mg = 40,000 IU of vitamin D₃ or vitamin D₂.

of vitamin D in the diet. Table 6 attempts to establish some reasonable estimates regarding safe dietary intakes of vitamin D_3 as a function of dietary exposure time for various species. In several of the species listed in Table 6 (the horse, chicken, turkey, and probably the Japanese quail), experiments have established that vitamin D_3 is 10 to 20 times more toxic than vitamin D_2. Therefore, the values in Table 6 should be adjusted accordingly for cases in which vitamin D_2 is the sole dietary source of vitamin D.

Very little information exists regarding the maximum safe dietary level of vitamin D_3 for a long (more than 60-day) exposure time. Horst and Littledike (1982) reported plasma vitamin D and vitamin D metabolite concentrations in several animal species that consumed experimental diets for several months. A retrospective analysis of dietary ingredients indicated that all of the diets consumed by the different species contained 4- to 10-fold the required level of vitamin D_3 (National Research Council, 1975, 1978a, 1979, 1984). Also, in these experiments, plasma 25-OH-D_3 concentrations, a sensitive indicator of vitamin D excess, were found to be within the normal range (20 to 80 ng/ml) for all of the different species included in the analysis. In sheep fed diets containing 10 times the level of required vitamin D according to the Agricultural Research Council, similar results have been obtained (Smith et al., 1985). The same workers, observed, however, that when dietary vitamin D_3 was 20-fold the sheep's nutritional requirement, plasma 25-OH-D_3 concentrations increased significantly. Most animal species appear to be able to tolerate 10 times the level of vitamin D that they require

for long periods of time. Catfish and rainbow trout, on the other hand, can tolerate as much as 20 and 500 times their requirements, respectively (Andrews et al., 1980; Hilton and Ferguson, 1982).

Under short-term feeding conditions (less than 60 days), most of the species listed in Table 6 can tolerate up to 100 times their apparent requirements for vitamin D. Experiments supporting this conclusion are, for the most part, extracted from Table 5.

Although most animals can tolerate excess vitamin D for extended periods, there has been no credible data suggesting that exceeding dietary requirements by several times improves performance. Therefore, other than to compensate for oxidative losses, there is no justification for feeding excessive dietary vitamin D.

CONCLUSION

More research is needed to further clarify the vitamin D mechanism that causes toxic effects in different species. Whether the tissue calcinosis is purely a result of hypercalcemia or due to some other factor is a question of prime importance. Also, there is little information regarding the quantity and distribution of vitamin D and vitamin D metabolites in affected tissues.

SUMMARY

1. Vitamin D is essential for normal bone formation in animals. It is required in the diets of animals raised with insufficient exposure to sunlight.

2. Studies indicate that vitamin D_3 is 10 to 20 times more toxic than vitamin D_2.

3. For most species the presumed maximal safe level of vitamin D_3 for long-term feeding conditions (more than 60 days) is 4 to 10 times the recognized dietary requirement. Under short-term feeding conditions (less than 60 days), most species can tolerate as much as 100 times their apparent dietary requirements.

4. There is no known benefit to feeding vitamin D to animals in excess of the recognized dietary requirement levels.

TABLE 6   Estimation of Safe Upper Dietary Levels of Vitamin D_3 for Animals

Species	Dietary Requirement[a]	Exposure Time < 60 d[b]	Exposure Time > 60 d
		IU/kg vitamin D_3 diet[c]	
Birds			
Chicken	200	40,000	2,800
Japanese quail	1,200	120,000	4,700
Turkey	900	90,000	3,500
Cow	300	25,000	2,200
Fish			
Catfish	1,000		20,000
Rainbow trout	1,800		1,000,000
Horse	400		2,200
Sheep	275	25,000	2,200
Swine	220	33,000	2,200

[a]From the National Research Council (1975, 1978a, 1978b, 1979, 1981, 1983, 1984).

[b]The safe upper level of vitamin D_3 for an exposure time of less than 60 days is undetermined for the horse, catfish, and rainbow trout.

[c]One IU = 0.025 μg of vitamin D_3.

REFERENCES

Andrews, J. W., T. Murai, and J. W. Page. 1980. Effects of dietary cholecalciferol and ergocalciferol on catfish. Aquaculture 19:49.

Capen, C. C., C. R. Cole, and J. W. Hibbs. 1966. The pathology of hypervitaminosis D in cattle. Pathol. Vet. 3:350.

Chen, P. S., and H. B. Bosmann. 1964. Effect of vitamin D_2 and D_3 on serum calcium and phosphorus in rachitic chicks. J. Nutr. 83:133.

Chineme, C. N., L. Krook, and W. G. Pond. 1976. Bone pathology in hypervitaminosis D. An experimental study in young pigs. Cornell Vet. 66:387.

Clegg, F. G., and J. G. Hollands. 1976. Cervical scoliosis and kidney lesions in sheep following dosage with vitamin D. Vet. Rec. 98:144.

Cole, C. R., D. M. Chamberlain, J. W. Hibbs, W. D. Pounden, and C. R. Smith. 1957. Vitamin D (Viosterol) poisoning in a cow, from a study on the prevention of parturient paresis. J. Am. Vet. Med. Assoc. 130:298.

DeLuca, H. F. 1971. Vitamin D: New horizons. Clin. Orthop. Relat. Res. 78:4.

DeLuca, H. F. 1984. The metabolism, physiology and function of vitamin D. Pp. 1–68 in Vitamin D: Basic and Clinical Aspects, R. Kumar, ed. Boston: Martimes Nijhoft Publishing.

Dinkel, L. 1966. Nephrocalcinosis and dichystrol (dihydrotachysterol) medication. Dtsch. Med. Wochenschr. 9:357.

Greig, W. A. 1963. Hypervitaminosis D in cattle. XVII World Vet. Congr., Hannover Proc. 1:233.

Harrington, D. D. 1982. Acute vitamin D_2 (ergocalciferol) toxicosis in horses: Case report and experimental studies. J. Am. Vet. Med. Assoc. 180:867.

Harrington, D. D., and E. H. Page. 1983. Acute vitamin D_3 toxicosis in horses: Case reports and experimental studies of the comparative toxicity of vitamins D_2 and D_3. J. Am. Vet. Med. Assoc. 182:1358.

Haussler, R. R., and T. A. McCain. 1977. Basic and clinical concepts related to vitamin D metabolism and action. N. Engl. J. Med. 297:974.

Helgebostad, A., and K. Nordstoga. 1978. Hypervitaminosis D in fur-bearing animals. Nord. Vet. 30:451.

Hibbs, J. W., and W. D. Pounden. 1955. Studies on milk fever in dairy cows. IV. Prevention by short-time prepartum feeding of massive doses of vitamin D. J. Dairy Sci. 38:65.

Hilton, J. W., and H. W. Ferguson. 1982. Effect of excess vitamin D_3 on calcium metabolism in rainbow trout salmon Gairdneri richardson. J. Fish Biol. 21:373.

Hines, T. G., N. L. Jacobson, D. C. Beitz, and E. T. Littledike. 1985. Dietary calcium and vitamin D. Risk factors in the development of atherosclerosis in the young goat. J. Nutr. 115:167.

Hintz, H. F., H. F. Schryver, J. E. Lowe, J. King, and L. Krook. 1973. Effect of vitamin D on Ca and P metabolism in ponies. J. Anim. Sci. 37:282.

Hollis, B. W., B. A. Roos, H. H. Draper, and P. W. Lambert. 1981. Vitamin D and its metabolites in human and bovine milk. J. Nutr. 111:1135.

Hollis, B. W., P. W. Lambert, and R. L. Horst. 1983. Factors affecting the antirachitic sterol content of native milk. P. 157 in Perinatal Calcium and Phosphorus Metabolism, M. F. Holick, T. K. Gray, and C. S. Anast, eds. New York: Elsevier.

Horst, R. L. 1986. Regulation of calcium and phosphorus homeostasis in the dairy cow. J. Dairy Sc. 69:604.

Horst, R. L., and E. T. Littledike. 1979. Elevated plasma 1,25-$(OH)_2$D following massive dosing of vitamin D_3 in dairy cattle. P. 999 in Vitamin D, Basic Research and Its Clinical Application. New York: Walter de Gruyter.

Horst, R. L., and E. T. Littledike. 1982. Comparison of plasma concentrations of vitamin D and its metabolites in young and aged domestic animals. Comp. Biochem. Physiol. 73B:485.

Horst, R. L., E. T. Littledike, J. L. Riley, and J. L. Napoli. 1981. Quantitation of vitamin D and its metabolites and their plasma concentrations in five species of animals. Anal. Biochem. 116:189.

Horst, R. L., J. L. Napoli, and E. T. Littledike. 1982. Discrimination in the metabolism of orally dosed ergocalciferol and cholecalciferol by the pig, rat and chick. Biochem. J. 204:185.

Horst, R. L., and T. A. Reinhardt. 1983. Vitamin D metabolism in ruminants and its relevance to the periparturient cow. J. Dairy Sci. 66:661

Hughes, M. R., D. J. Baylink, W. A. Gonnerman, S. V. Toverund, W. K. Ramp, and M. R. Haussler. 1977a. Influence of dietary vitamin D_3 on the circulating concentration of its active metabolites in chick and rat. Endocrinology 100:799.

Hughes, M. R., T. A. McCain, S. Y. Change, M. R. Haussler, M. Villareale, and R. H. Wasserman. 1977b. Presence of 1,25-dihydroxyvitamin D_3-glycoside in the calcinogenic plant Cestrum diurnum. Nature 269:347.

Hunt, R. D., F. G. Garcia, D. M. Hegsted, and N. Kaplinsky. 1967. Vitamin D_2 and vitamin D_3 in new world primates: Influence on calcium absorption. Science 157:943.

Hunt, R. D., F. G. Garcia, and D. M. Hegsted. 1969. Hypervitaminosis D in new world monkeys. Am. J. Clin. Nutr. 22:358.

Hunt, R. D., F. G. Garcia, and R. J. Walsh. 1972. A comparison of the toxicity of ergocalciferol and cholecalciferol in rhesus monkeys. J. Nutr. 102:975.

Kamio, A., F. A. Kummerow, and H. Imai. 1977. Degeneration of aortic smooth muscle cells in swine fed excess vitamin D_3. Arch. Pathol. Lab. Med. 101:378.

Kent, S. P., G. F. Vawter, R. M. Dowben, and R. E. Benson. 1958. Hypervitaminosis D in monkeys; a clinical and pathological study. Am. J. Pathol. 34:37.

Krauss, W. E., and R. M. Bethke. 1937. New developments in the field of vitamin D milk. Ohio Agric. Exp. Stn. Bull. 184:3.

Kunz, C., M. Niesen, H. V. Lilienfeld-Toal, and W. Burmeister. 1984. Vitamin D, 25-hydroxy-vitamin D and 1,25-dihydroxy-vitamin D in cow's milk, infant formula and breast milk during different stages of lactation. Int. J. Vit. Nutr. Res. 54:141.

Littledike, E. T., and R. L. Horst. 1982. Vitamin D_3 toxicity in dairy cows. J. Dairy Sci. 65:749.

Long, G. G. 1984. Acute toxicosis in swine associated with excessive dietary intake of vitamin D. J. Am. Vet. Med. Assoc. 184:164.

Manston, R., and J. M. Payne. 1964. Mineral imbalance in pregnant "milk fever prone" cows and the value and possible toxic side effects of treatment with vitamin D_3 and dihydrotachysterol. Br. Vet. J. 120:167.

McDermott, C. M., D. C. Beitz, E. T. Littledike, and R. L. Horst. 1985. Effects of dietary vitamin D_3 on concentrations of vitamin D and its metabolites in blood plasma and milk of dairy cows. J. Dairy Sci. 68:1,959.

Metz, A. L., M. M. Wolser, and W. G. Olson. 1985. The interaction of vitamin A and vitamin D related to skeletal development in turkey poultry. J. Nutr. 115:929.

Morrissey, R. L., R. M. Cohen, R. N. Empson, H. L. Greene, O. D. Taunton, and Z. Z. Ziporin. 1977. Relative toxicity and metabolic effects of cholecalciferol and 25-hydroxycholecalciferol in chicks. J. Nutr. 107:1027.

Mullan, P. A., P. G. C. Bedford, and P. L. Ingram. 1979. An investigation of the toxicity of 1α-hydroxycholecalciferol to calves. Res. Vet. Sci. 27:275.

Napoli, J. L., J. L. Sommerfeldt, B. C. Pramanik, R. Gardner, A. D. Sherry, J. J. Partridge, M. R. Uskokovic, and R. L. Horst. 1983. 19-Nor-10-keto-vitamin D derivatives: Unique metabolites of vitamin D_3, vitamin D_2 and 25-hydroxyvitamin D_3. Biochemistry 22:3636.

National Research Council. 1975. Nutrient Requirements of Sheep. 5th rev. ed. Washington, D.C.: National Academy Press.

National Research Council. 1978a. Nutrient Requirements of Dairy Cattle. 5th rev. ed. Washington, D.C.: National Academy Press.

National Research Council. 1978b. Nutrient Requirements of Horses. 4th rev. ed. Washington, D.C.: National Academy Press.

National Research Council. 1979. Nutrient Requirements of Swine. 8th rev. ed. Washington, D.C.: National Academy Press.

National Research Council. 1981. Nutrient Requirements of Coldwater Fishes. Washington, D.C.: National Academy Press.

National Research Council. 1983. Nutrient Requirements of Warmwater Fishes and Shellfishes. Rev. ed. Washington, D.C.: National Academy Press.

National Research Council. 1984. Nutrient Requirements of Poultry. 8th rev. ed. Washington, D.C.: National Academy Press.

Norman, A. W. 1979. Vitamin D. In The Calcium Homeostatic Hormone, W. J. Darby, ed. New York: Associated Press.

Norman, A. W. 1980. 1,25-Dihydroxyvitamin D_3 and 24,25-dihydroxyvitamin D_3: Key components of the vitamin D endocrine system. Contemp. Nephrol. 18:1.

Norman, A. W., J. Roth, and L. Orci. 1982. The vitamin D endocrine system—Steroid metabolism, hormone receptors, and biological response (calcium binding proteins). Endocrine Rev. 3:331.

Olson, W. G., N. A. Jorgensen, A. N. Bringe, L. H. Schultz, and H. F. DeLuca. 1973. 25-Hydroxycholecalciferol (25-OH-D_3). III. Effect of dosage on soft tissue integrity and vitamin D activity of tissue and milk from dairy cows. J. Dairy Sci. 57:677.

Parfitt, A. M. 1977. Renal function in treated hypoparathyroidism, a possible direct nephrotoxic effect of vitamin D. Adv. Exp. Med. Biol. 81:455.

Parfitt, A. M. 1980. Surgical, idiopathic and other varieties of parathyroid hormone-deficient hypoparathyroidism. Pp. 755–768 in Metabolic Basis of Endocrinology, L. Degroot, ed. New York: Grune & Stratton.

Payne, J. M., and R. Manston. 1967. The safety of massive doses of vitamin D_3 in the prevention of milk fever. Vet. Rec. 79:214.

Poston, H. A. 1968. Effects of massive doses of vitamin D_3 on fingerling brook trout. Fish. Res. Bull. 32:48.

Putscher, W. 1929. Über Vigantolschädigung der Niere einum Kinde. Z. Kinderheilkd. 48:269.

Quarterman, J., A. C. Dalgarna, A. Adams, B. F. Fell, and R. Boyne. 1964. The distribution of vitamin D between the blood and liver in the pig, and observations on the pathology of vitamin D toxicity. Br. J. Nutr. 18:65.

Ratzkowski, C., N. Fine, and S. Edelstein. 1982. Metabolism of cholecalciferol in vitamin D intoxicated chicks. Isr. J. Med. Sci. 18:695.

Recker, R. R., P. Schoenfeld, J. Letteri, E. Slatopolsky, R. Goldsmith, and A. Brickman. 1979. 25-Hydroxyvitamin D (calcidiol) in renal osteodystrophy: Long term results of a six center trial. P. 869 in Vitamin D: Basic Research and Its Clinical Application, A. W. Norman, K. Schaefer, D. V. Herrath, H. G. Grigolet, J. W. Coburn, H. F. DeLuca, E. B. Mawer, and T. Suda, eds. Berlin: Walter de Gruyter.

Reeve, L. E., N. A. Jorgensen, and H. F. Deluca. 1983. Vitamin D compounds in cow's milk. J. Nutr. 112:667.

Seekles, L., N. C. W. Hesse, and J. B. van Dijk. 1961. The prevention of undesirable side effects in the preventive treatment of milk fever by means of parenteral administration of solubilized vitamin D_3. Tijdschr. Diergeneeskd. 5:344.

Simesen, M. G., T. Hanichen, and K. Dammrich. 1978. Hypervitaminosis D in sheep. Acta Vet. Scand. 19:588.

Smith, B. S. W., H. Wright, G. U. Aitchison, and J. A. Spence. 1985. Effects in sheep of high dietary levels of vitamin D_3 as measured by circulating concentrations of some vitamin D metabolites and skeletal examination. Res. Vet. Sci. 18:317.

Soares, J. H., D. M. Kaetzel, J. T. Allen, and M. R. Swerdel. 1982. Toxicity of a vitamin D steroid in laying hens. Poult. Sci. 62:24.

Spangler, W. L., D. H. Gribble, and T. C. Lee. 1979. Vitamin D intoxication and the pathogenesis of vitamin D nephropathy in the dog. Am. J. Vet. Res. 40:73.

Stern, P. H. 1980. The D vitamins and bone. Pharmacol. Rev. 32:47.

Stevens, V. I., and R. Blair. 1985. Effects of supplemental vitamin D_3 on egg production of two strains of Japanese quail and growth of their progeny. Poult. Sci. 64:510.

Streck, W. F., C. Waterhouse, and J. G. Haddad. 1979. Glucocorticoid effects in vitamin D intoxication. Arch. Intern. Med. 139:974.

Taylor, T. G., K. M. L. Morris, and J. Kirkley. 1968. Effects of dietary excesses of vitamins A and D on some constituents of the blood of chicks. Br. J. Nutr. 22:713.

Toda, T., D. E. Leszczynski, and F. A. Kummerow. 1983. The role of 25-hydroxyvitamin D_3 in the induction of atherosclerosis in swine and rabbit by hypervitaminosis. Acta Pathol. Jpn. 33:37.

Ulbrych-Jablonska, A. 1972. The hypocalcemic effect of glucagon in cases of hypercalcemia. Helv. Paediatr. Acta 27:613.

West, T. E. T., M. Joffe, L. Sinclair, and J. L. H. O'Riordan. 1971. Treatment of hypercalcemia with calcitonin. Lancet 1:675.

Wilson, J. H. G. 1964. On the milk fever syndrome in the bovine. Ph.D. dissertation. Rijksuniversiteit te Utrecht, Netherlands.

Wren, W. B. 1980. Hypervitaminosis D (vitamin D toxicosis) in weanling pigs. Proc. Annu. Mtg. Am. Assoc. Vet. Lab. Diagnost. 23:101.

Vitamin E

Vitamin E was recognized more than 60 years ago as a factor required for normal gestation in rats fed diets containing rancid fat (Evans and Bishop, 1922). This factor, named tocopherol from the Greek *tokos* (childbirth) and *pherein* (to bring forth), was also found to be required for prevention of encephalomalacia in chicks and nutritional myopathies in several species (Goettsch and Pappenheimer, 1931; Pappenheimer and Goettsch, 1931). Evans et al. (1936) isolated the vitamin from wheat germ oil; Fernholz (1938) elucidated its chemical structure; and Karrer et al. (1938) achieved its synthesis shortly thereafter.

NUTRITIONAL ROLE

Dietary Requirements of Various Species

After vitamin E was recognized as an essential nutrient, numerous interrelationships were identified between it and other dietary factors, such as selenium and synthetic antioxidants, in preventing many varied animal diseases. (See reviews by Mason and Horwitt, 1972; Scott, 1978; Combs, 1981; Machlin, 1980, 1984.) These diseases include those prevented by vitamin E or certain synthetic antioxidants (e.g., encephalomalacia in chicks, fetal death and resorption in rats, depigmentation of incisor enamel in rats, and muscular dystrophy in rabbits); those prevented by vitamin E or selenium (e.g., dietary liver degeneration in rats, exudative diathesis in chicks, and nutritional muscular dystrophies in lambs, calves, ducks, and turkeys); and those prevented only by vitamin E (e.g., testicular degeneration in rats, hamsters, guinea pigs, dogs, monkeys, and chickens, and nutritional muscular dystrophies in rats, guinea pigs, rabbits, pigs, and dogs). The dietary requirements for vitamin E estimated for most animal species are in the range of 5 to 50 IU/kg of diet. The role of vitamin E in human health is most apparent in conditions of poor enteric absorption of lipids, for example, biliary atresia, cystic fibrosis, and neonatal prematurity. Similar conditions of lipid malabsorption in animals, such as pancreatitis or bile stasis, may be expected to impair the utilization of dietary vitamin E.

Biochemical Functions

Because synthetic antioxidants, such as ethoxyquin, diphenyl-*p*-phenylenediamine (DPPD), and butylated hydroxytoluene (BHT) prevent many vitamin E-deficiency syndromes and because vitamin E functions in vitro as a very good antioxidant, hypotheses for this nutrient's mode of action held that it was a biologically specific lipid-soluble antioxidant (Tappel, 1962). However, the metabolic basis for the nutritional interrelationships of vitamin E and selenium was not understood until Rotruck et al. (1972) discovered that selenium was an essential component of an enzyme, glutathione peroxidase, which was involved in the metabolism of hydroperoxides. Investigations of this interrelationship have led to the present understanding that vitamin E and selenium (via glutathione peroxidase) function as parts of a multicomponent antioxidant defense system. This system protects the cell against the adverse effects of reactive oxygen and other free radical initiators of the oxidation of polyunsaturated membrane phospholipids, critical proteins, or both (Chow, 1979). This function of vitamin E is thought to be the basis of its role in nutrition and in protection against the toxic effects of certain pro-oxidant drugs (Combs, 1981).

The different types of vitamin E-deficiency syndromes that are manifested in different animals have been taken to indicate that, in various species and organ systems, lesions in different aspects of the cellular anti-

TABLE 8 Research Findings of High Levels of Vitamin E in Animals

Species and No. of Animal	Age or Weight	Administration Amount	Form	Duration	Route	Effect	Reference
Chickens, 4 groups of 6	37 d	2,200 IU/kg diet	All-*rac*-α-tocopheryl acetate	23–27 d, posthatching	Diet	Depressed growth rate; reduced hematocrit; reticulocytosis; increased prothrombin time (corrected by injecting vitamin K); bone calcification depressed when chicks fed calcium- or vitamin D-deficient diet; skeletal muscle mitochondria showed 33% reduction in oxygen uptake; requirement for vitamin D and K increased	March et al., 1973
Chickens, 3 groups of 10	21 d	10,000 IU/kg diet	All-*rac*-α-tocopheryl acetate	18–21 d, posthatching	Diet	Reduced calcium and phosphorus in plasma and dry fat-free bone ash (femur)	Murphy et al., 1981
Chickens, 10	35 d	64,000 IU/kg diet	All-*rac*-α-tocopheryl acetate	1–35 d, posthatching	Diet	Hepatomegaly; reduced pigmentation in beak, feet, and shanks; reduced chick BW; waxy-appearing feathers	Nockels et al., 1976
Chickens, 6	25 d	200 IU/kg diet	All-*rac*-α-tocopheryl acetate	21–25 d, posthatching	Diet	High vitamin E stores and decreased intraduodenal concentrations of retinyl glucuronides with no effect on enteric absorption of vitamin A	Sklan, 1983
Rats, 9	457 g	250 IU/kg diet	All-*rac*-α-tocopheryl acetate	16 mo	Diet	Decrease in femur ash content; increase in plasma alkaline phosphatase activity and relative heart and spleen weights; elevated hematocrit value	Yang and Desai, 1977a

Species, no.	BW	Dose	Compound	Duration	Route	Effect	Reference
Rats, 5	358 g	25,000 IU/kg diet	All-*rac*-α-tocopheryl acetate	16 mo	Diet	Decrease in femur ash content; increase in plasma alkaline phosphatase activity and relative heart and spleen weights; elevated hematocrit value	Yang and Desai, 1977a
Rats, 4		25,000 IU/kg diet	All-*rac*-α-tocopheryl acetate	16 mo	Diet	Vitamin E accumulated in liver, proportional to dietary intake; plasma total lipids and cholesterol lowered	Yang and Desai, 1977b
Rats, 5	85–96 g	2,000 IU/kg diet	All-*rac*-α-tocopheryl acetate	2 wk	Diet	Plasma triglycerides and phospholipids significantly increased	Cho and Sugano, 1978
Rats, 10	384 g	6,000 IU/kg diet	All-*rac*-α-tocopheryl acetate	8 wk	Diet	Significant reduction in relative weight of adrenal glands; significant increase in liver retinyl ester concentration; increase in plasma albumin concentration; decrease in plasma globulin concentration resulting in 50 percent increase in A:G ratio	Jenkins and Mitchell, 1975
Rats, 6	180 g	2,500 IU/kg diet	2-*ambo*-α-tocopherol	In utero-48 d	Diet	Deleterious effect on percent ash or mineral composition of developing rat teeth; molars had slightly higher calcium and phosphorus levels	Alam and Alam, 1981
Rats, 60	134 g initial BW	2,000 IU/kg BW/day	All-*rac*-α-tocopheryl acetate	104 wk	Diet	No adverse affect on growth rate, survival, or hepatic function	Weldon et al. 1983
Rats, 11	210 g initial BW	2,252 IU/kg BW/day	All-*rac*-α-tocopheryl acetate	Gestation and lactation	Diet	At term, significantly higher plasma vitamin E concentrations as compared to controls; no teratogenic effects observed; a few delayed deliveries (21 d) and a few pups with eyes closed at 14 d of lactation noted	Martin and Hurley 1977
Rats, mice, rabbits		Graded doses acute oral LD$_{50}$	All-*rac*-α-tocopheryl acetate		Oral	Oral values estimated at >2 g/kg BW for each species	FASEB 1975

thrombin times were observed. Vitamin K injections corrected the prothrombin times. The high level of vitamin E depressed bone calcification among chicks fed either a calcium- or vitamin D-deficient diet. Skeletal muscle mitochondria isolated from chicks fed the high level of vitamin E showed a 33 percent reduction in oxygen uptake. March et al. (1973) concluded that vitamin E fed to chicks at the 2,200-IU/kg level increased their nutritional requirements for vitamins K and D. Murphy et al. (1981) showed an effect related to the vitamin D function. Their research found that vitamin E at 10,000 IU/kg of diet reduced concentrations of calcium and phosphorus in plasma and of total ash in tibiae.

Nockels et al. (1976) found that dietary levels of vitamin E of 4,000 IU/kg or more produced hepatomegaly and reduced skin pigmentation in broiler chicks. Levels of 8,000 IU/kg or more significantly reduced chick body weight (BW) and caused a waxy appearance of the feathers. Sklan (1983) found that vitamin E at 200 IU/kg of diet increased hepatic vitamin A stores and decreased intraduodenal concentrations of retinyl glucuronides with no effect on the enteric absorption of vitamin A.

Yang and Desai (1977a,b) conducted long-term studies of the effects of high dietary levels of vitamin E (all-*rac*-α-tocopheryl acetate) on growth in rats. It was evident by 8 months that levels of vitamin E of 10,000 IU/kg significantly depressed BW and increased relative heart weights (organ weight/unit BW) and by 16 months that relative spleen weights increased. That level of vitamin E also depressed femur ash content by 16 months and decreased prothrombin times at 12 months. Hematocrit values were increased at 12 and 16 months in rats fed 25,000 IU of vitamin E/kg. Rats fed 2,500 IU of vitamin E/kg showed increased hepatic lipid contents at 8 months, but this effect was not significant at 16 months. Rats fed 10,000 or 25,000 IU of vitamin E/kg showed reductions in the total lipids and cholesterol contents of plasma by 16 months. This finding contrasts with the report of Cho and Sugano (1978), who found that a dietary level of 2,000 IU of vitamin E/kg tended to cause higher plasma lipid levels in the rat.

Yang and Desai (1977a,b) found that high-level vitamin E treatment did not significantly affect liver vitamin A storage or urinary creatine or creatinine. Jenkins and Mitchell (1975), however, found that a dietary level of 6,000 IU of vitamin E/kg produced significant increases in liver retinyl ester concentrations, both at low and intoxicating levels of vitamin A intake. High levels of vitamin E have been shown to reduce the hepatic storage of vitamin A (Johnson and Baumann, 1948; Swick and Baumann, 1951).

Jenkins and Mitchell (1975) found that a dietary vitamin E level of 6,000 IU/kg did not affect the 8-week growth of weanling rats. This level of the vitamin significantly reduced the relative weight of the adrenal gland but did not affect the relative weights of liver, kidney, spleen, or testes. Although the total protein concentration of plasma was not significantly affected, the high level of vitamin E increased albumin concentrations and decreased globulin concentrations. The result was a 50 percent increase in the albumin: globulin ratio.

Yang and Desai (1977a,b) observed no adverse effects of any kind among rats fed levels of vitamin E as great as 2,500 IU/kg. Alam and Alam (1981) found the same dietary level of vitamin E to produce no deleterious effects on ash or mineral contents of developing rat teeth. Wheldon et al. (1983) found that dietary intakes of vitamin E as great as 2,000 mg of all-*rac*-α-tocopheryl acetate/kg of BW/day for 104 weeks did not adversely affect growth rate, survival, or hepatic function as indicated by serum enzyme levels.

Martin and Hurley (1977) studied the effects of excessive amounts of vitamin E during pregnancy and lactation in the rat. They found that the placental transfer of vitamin E is inefficient; thus, the dietary exposure of the dams to vitamin E had minimal effects on the progeny before birth. They observed no teratogenic effects of dietary intakes as great as 2,252 mg/kg of BW per day; however, this level of vitamin E intake was associated with a few cases of delayed deliveries (i.e., gestation periods longer than 21 days) and a few pups with eyes closed at 14 days of age. The dams receiving the high level of vitamin E had enlarged livers and elevated plasma lipids.

The acute oral LD$_{50}$ value of all-*rac*-α-tocopheryl acetate for rats, mice, and rabbits has been estimated to be in excess of 2 g/kg of BW (FASEB, 1975).

Alberts et al. (1978) found that intraperitoneal administration of 85 IU of vitamin E to mice 24 hours before intravenous treatment with adriamycin increased the bone marrow toxicity of the drug.

Farrell and Bieri (1975) studied a population of 28 adults who consumed 100 to 800 IU of vitamin E/day for an average of 3 years. The results of clinical blood tests revealed no disturbances in the liver, kidney, muscle, thyroid gland, erythrocytes, leukocytes, coagulation parameters, and blood glucose. Farrell and Bieri concluded that vitamin E in this range of intake produced no apparent toxic side effects. Nevertheless, the literature contains reports of such effects as creatinuria (Hillman, 1957), fatigue (Roberts, 1981), depression (Kligman, 1982), thrombophlebitis (Roberts, 1978, 1981), and other disorders ranging from hypoglycemia to hypertension (Roberts, 1981). A review by Salkeld (1979) of more than 10,000 cases in which the minimum oral intake of vitamin E was greater than 200 IU/day for at least 4 weeks indicated that only 61 subjects reported side effects. These effects were generally minor: nausea, generalized dermatitis, and fatigue.

Tsai et al. (1978) conducted a double-blind study with 200 healthy college students who were given either 600 IU of vitamin E/day or a placebo. Their results showed that vitamin E treatment did not significantly affect subjective evaluations of work performance, sexuality, general well-being, muscular weakness, or gastrointestinal disturbances. It also did not affect prothrombin times, total blood leukocyte counts, or serum creatine phosphokinase activities. Vitamin E treatment did produce significant elevations in serum triglycerides in females. It significantly decreased serum concentrations of thyroxine and triiodothyronine in females who were not using oral steroid contraceptive agents and in males.

Corrigan and Marcus (1974) reported a coagulopathy, which is characterized by severely prolonged prothrombin times, in a patient receiving anticoagulant therapy and voluntarily consuming a high level (1,200 IU/day) of vitamin E. A model for this condition has been produced in the dog (Corrigan, 1979). He showed that high levels of vitamin E do not affect coagulation mechanisms unless animals are made mildly vitamin K deficient by the use of warfarin. In this case, high levels of vitamin E produce a profound coagulopathy. A double-blind study by Zipursky et al. (1980) found that administration of 25 IU/day of vitamin E by mouth to premature infants to 6 weeks of age did not affect coagulation factors.

PRESUMED UPPER SAFE LEVELS

For the time being, the information on hypervitaminosis E in animals is limited. Therefore, estimates of maximum tolerable levels in animals should be considered tentative. Studies with rats and chicks indicate that dietary levels of at least 1,000 IU/kg can be fed for prolonged periods of time without deleterious effects. For these species, the presumed upper safe levels of vitamin E are higher than the dietary levels by rather undefined increments. In rats, the maximum tolerable level is probably about 2,500 IU/kg. The studies by Yang and Desai (1977a,b) and Alam and Alam (1981) indicate that this level is not hazardous. The presumed upper safe level for the chick, however, is lower (1,000 to 2,000 IU/kg) as indicated by the studies of March et al. (1973). The level of 1,000 IU/kg is taken, therefore, as the presumed upper safe level of vitamin E for the chick. In the absence of experimental data on hypervitaminosis E for other species, maximum tolerable levels of the vitamin can be inferred only by extrapolation from these estimates for rats and chicks. Thus, a presumed upper safe level of about 75 IU/kg of BW/day is suggested as a tentative guideline for safe dietary exposure to vitamin E. Because the dietary requirements of most species for vitamin E are in the range of 5 to 50 IU/kg of diet (or 2 to 4 IU/kg of BW/day), intakes of at least 20 times the nutritionally adequate levels should be well tolerated.

SUMMARY

1. Vitamin E is a required nutrient for cell antioxidant protection by all animals.

2. Hypervitaminosis E has been studied in rats, chicks, and humans. These scant data indicate maximum tolerable levels to be in the range of 1,000 to 2,000 IU/kg diet. A tentative presumed safe use level of 75 IU/kg of BW/day is suggested.

REFERENCES

Alam, S. Q., and B. S. Alam. 1981. Effects of excess vitamin E on rat teeth. Calcif. Tissue Int. 33:619.

Alberts, D. S., Y. M. Peng, and T. E. Moon. 1978. Alpha-tocopherol pretreatment increases adriamycin bone toxicity. Biomedicine 29:189.

Cho, S., and M. Sugano. 1978. Effect of different levels of dietary alpha tocopherol and linoleate on plasma and liver lipids in rats. J. Nutr. Sci. Vitaminol. 24:221.

Chow, C. K. 1979. Nutritional influence on cellular antioxidant defense systems. Am. J. Clin. Nutr. 32:1066.

Combs, G. F., Jr. 1981. Assessment of vitamin E status in animals and man. Proc. Nutr. Soc. 40:187.

Corrigan, J. J., Jr. 1979. Coagulation problems relating to vitamin E. Am. J. Pediat. Hematol. Oncol. 1:169.

Corrigan, J. J., Jr., and F. I. Marcus. 1974. Coagulapathy associated with vitamin E ingestion. J. Am. Med. Assoc. 230:1300.

Evans, H. M., and K. S. Bishop. 1922. On the existence of a hitherto unrecognized dietary factor essential for reproduction. Science (N.Y.) 56:650.

Evans, H. M., O. H. Emerson, and G. A. Emerson. 1936. The isolation from wheat-germ oil of an alcohol, α-tocopherol, having the properties of vitamin E. J. Biol. Chem. 113:319.

Farrell, P. M., and J. G. Bieri. 1975. Megavitamin E supplementation in man. Am. J. Clin. Nutr. 28:1381.

FASEB. 1975. Evaluations of the health aspects of tocopherols and alpha-tocopheryl acetate as food ingredients. Rep. No. PB 262 653. Bethesda, Md.: Federation of American Societies for Experimental Biology and Medicine.

Fernholz, E. 1938. Constitution of α-tocopherol. J. Am. Chem. Soc. 60:700.

Gallo-Torres, H. E. 1980a. Absorption. Pp. 170–172 in Vitamin E: A Comprehensive Treatise, L. J. Machlin, ed. New York: Marcel Dekker.

Gallo-Torres, H. E. 1980b. Blood transport and metabolism. Pp. 193–267 in Vitamin E: A Comprehensive Treatise, L. J. Machlin, ed. New York: Marcel Dekker.

Goettsch, M., and A. M. Pappenheimer. 1931. Nutritional muscular dystrophy in the guinea pig and rabbit. J. Exp. Med. 54:145.

Hillman, R. W. 1957. Tocopherol excess in man: Creatinuria with prolonged ingestion. Am. J. Clin. Nutr. 5:597.

Jenkins, M. Y., and G. V. Mitchell. 1975. Influence of excess vitamin E on vitamin A toxicity in rats. J. Nutr. 105:1600.

Johnson, R. M., and C. A. Baumann. 1948. The effect of alpha-tocopherol on the utilization of carotene by the rat. J. Biol. Chem. 175:811.

Karrer, P., H. Fritzsche, B. H. Ringer, and N. J. Salomen. 1938. Synthesis of alpha-tocopherol (vitamin E). Nature 141:1057.

Kligman, A. M. 1982. Vitamin E toxicity. Arch. Dermatol. 118:289.

Machlin, L. J., ed. 1980. Vitamin E: A Comprehensive Treatise. New York: Marcel Dekker.

Machlin, L. J. 1984. Vitamin E. Pp. 99–145 in Handbook of Vitamins: Nutritional, Biochemical, and Clinical Aspects, L. J. Machlin, ed. New York: Marcel Dekker.

March, B. E., E. Wong, L. Seier, J. Sim, and J. Biely. 1973. Hypervitaminosis E in the chick. J. Nutr. 103:371.

Martin, M. M., and L. S. Hurley. 1977. Effect of large amounts of vitamin E during pregnancy and lactation. Am. J. Clin. Nutr. 30:1629.

Mason, K. E., and M. K. Horwitt. 1972. Tocopherols. X. Effects of deficiency in animals. Pp. 272–292 in The Vitamins: Chemistry, Physiology, Pathology, Methods, Vol. 5, 2nd ed., W. H. Sebrell, Jr., and R. S. Harris, eds. New York: Academic Press.

Murphy, T. P., K. E. Wright, and W. J. Pudelkiwicz. 1981. An apparent rachitogenic effect of excessive vitamin E intakes in the chick. Poult. Sci. 60:1873.

Nockels, C. F., D. L. Menge, and E. W. Kienholz. 1976. Effect of excessive dietary vitamin E in the chick. Poult. Sci. 55:649.

Pappenheimer, A. M., and M. Goettsch. 1931. Cerebellar disorder in chicks, apparently of nutritional origin. J. Exp. Med. 53:11.

Roberts, H. J. 1978. Vitamin E and thrombophlebitis. Lancet 1:49.

Roberts, H. J. 1981. Perspective on vitamin E therapy. J. Am. Med. Assoc. 246:129.

Rotruck, J. T., A. L. Pope, H. E. Ganther, A. B. Swanson, D. G. Hafeman, and W. G. Hoekstra. 1972. Selenium: Biochemical role as a component of glutathione peroxidase. Science (N.Y.) 179:588.

Salkeld, R. M. 1979. Safety and tolerance of high-dose vitamin E administration in man: A review of the literature. Fed. Regist. 44:16172.

Scott, M. L. 1978. Vitamin E. Pp. 133–210 in The Lipid Soluble Vitamins, H. F. DeLuca, ed. New York: Plenum.

Sklan, D. 1983. Vitamin A absorption and metabolism in the chick: response to high dietary intake and to tocopherol. Br. J. Nutr. 50:401.

Swick, R. W., and C. A. Baumann. 1951. Effect of certain tocopherols and other antioxidants on the utilization of beta-carotene for vitamin A storage. Arch. Biochem. Biophys. 36:120.

Tappel, A. L. 1962. Vitamin E as the biological lipid antioxidant. Vit. Horm. (N.Y.) 20:493.

Tsai, A. C., J. J. Kelley, B. Peng, and N. Cook. 1978. Study on the effect of megavitamin E supplementation in man. Am. J. Clin. Nutr. 31:831.

Wheldon, G. H., A. Bhatt, P. Keller, and H. Hummler. 1983. D,L-alpha-tocopheryl acetate (vitamin E): A long-term toxicity and carcinogenicity study in rats. Int. J. Vit. Nutr. Res. 53:287.

Yang, N. Y. J., and I. D. Desai. 1977a. Effect of high levels of dietary vitamin E on hematological indices and biochemical parameters in rats. J. Nutr. 107:1410.

Yang, N. Y. J., and I. D. Desai. 1977b. Effect of high levels of dietary vitamin E on liver and plasma lipids and fat soluble vitamins in rats. J. Nutr. 107:1418.

Zipursky, A., R. A. Miller, V. S. Blanchette, and M. A. Johnston. 1980. Effect of vitamin E therapy on blood coagulation in newborn infants. Pediatrics 66:547.

Vitamin K

Henrik Dam discovered vitamin K as the result of experiments he was carrying out to determine whether cholesterol was a dietary essential. Dam (1929) noted a hemorrhagic syndrome in chicks fed diets that had sterols extracted by lipid solvents. He eventually isolated an active antihemorrhagic factor from alfalfa that E. A. Doisy's research group characterized as 2-methyl-3-phytyl-1,4-napthoquinone (MacCorquodale et al., 1939). Early investigations established that vitamin K deficiency resulted in decreased activity of prothrombin in plasma. As they were discovered, the synthesis of clotting factors VII, IX, and X was also shown to be vitamin K-dependent. More recently, it has been shown that a number of other proteins, many with presently undetermined functions, require vitamin K for their biosynthesis. Dam (1964), Almquist (1975), and Suttie (1985a) have reviewed the historical and more recent aspects of vitamin K nutrition and function.

NUTRITIONAL ROLE

Dietary Requirements of Various Species

Dietary adequacy of vitamin K is often defined as the amount of vitamin needed to maintain normal levels of plasma vitamin K-dependent clotting factors. It has been difficult to demonstrate clearly and to measure dietary vitamin K requirements for many species. Presumably this is because of the varying degrees to which different species utilize the large amount of vitamin K synthesized by bacteria in the lower gut and the degree to which they practice coprophagy. A spontaneous deficiency of vitamin K was first noted in chicks; poultry are much more likely to develop dietary deficiency signs than any other species. Ruminants do not appear to need a source of vitamin K in the diet because the vita-

min is synthesized by rumen microorganisms and subsequently utilized. Vitamin K deficiencies have been produced in most nonruminant species. Nevertheless, estimations of requirements made by different workers are difficult to compare because of the use of different forms of the vitamin and different methods to establish dietary requirements. Some investigators have measured the amount of vitamin K needed to cure the clotting defect in vitamin K-deficient hypoprothrombinemic animals; others have determined the minimal dietary concentrations of vitamin needed to prevent appearance of hypoprothrombinemia.

Scott (1966), Doisy and Matschiner (1970), and Griminger (1971) have discussed the vitamin K requirements of various species in detail. The requirement for most species ranges from 1 to 10 μg of vitamin K/kg of BW/day (50 to 150 μg/kg of diet) for most nonruminant animals; 50 to 250 μg/kg of BW/day (0.5 to 1.5 mg/kg of diet) for poultry. A dietary vitamin K requirement in the adult human has been difficult to establish but is usually stated to range from 0.5 to 1.5 μg vitamin K/kg of BW/day (Olson, 1980).

Biochemical Functions

In the absence of adequate amounts of vitamin K or in the presence of vitamin K antagonists, animals develop bleeding disorders. These disorders result from an inability of a liver microsomal enzyme, now called the vitamin K-dependent carboxylase (Esmon et al., 1975), to carry out the normal post-translational conversion of specific glutamyl residues in the vitamin K-dependent plasma proteins to γ-carboxyglutamyl residues (Nelsestuen et al., 1974; Stenflo et al., 1974). These residues are essential for the normal Ca^{++}-dependent interaction of the vitamin K-dependent clotting factors with phospholipid surfaces. The result of insufficient vitamin K to

31

serve as a cofactor for this enzyme is, therefore, a decrease in the rate of thrombin generation. This decrease subsequently results in a decreased rate of fibrin clot formation and an increased susceptibility to hemorrhage. The molecular role of vitamin K in the enzymatic reaction is understood in a general sense, but a number of details of the chemical mechanisms remain to be clarified (Suttie, 1985b).

FORMS OF THE VITAMIN

Compounds exhibiting vitamin K activity possess a 2-methyl-1,4- napthoquinone ring. The term "vitamin K" is used as a generic descriptor for both 2-methyl-1,4-napthoquinone and all 3-substituted derivatives of this compound, which exhibit an antihemorrhagic activity in animals fed a vitamin K-deficient diet (see Figure 8). The form of the vitamin isolated from plants, 2-methyl-3-phytyl-1,4-napthoquinone, is generally called vitamin K_1 or phylloquinone. The vitamin originally isolated from purified fish meal, which was called vitamin K_2, is now known to be only one of a series of bacterially synthesized vitamins K with unsaturated polyisoprenoid side chains at the 3-position. These are called menaquinones (MK). The predominant vitamins of the menaquinone series contain a side chain of 6 to 10 isoprenoid units (MK-6 through MK-10); however, forms with up to 13 isoprenoid groups have been identified, as well as several forms with partially saturated side chains. The parent compound of the vitamin K series, 2-methyl-1, 4-napthoquinone, is a synthetic product that was once called vitamin K_3 but is now more commonly and correctly designated as menadione. Enzymes in mammalian tissue are capable of alkylating menadione to active forms of the vitamin, and MK-4 appears to be the predominant species formed.

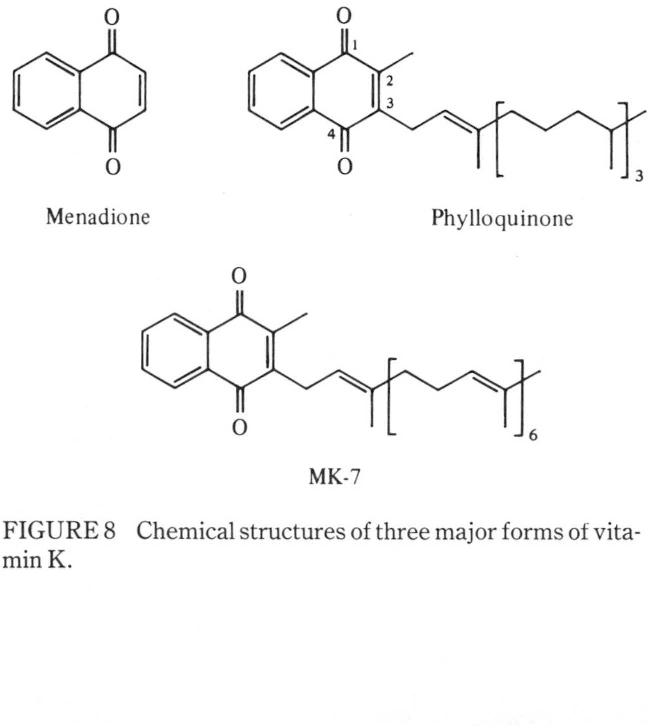

A limited number of forms of vitamin K are currently available for therapeutic and nutritional use. Phylloquinone (USP phytonadione) is the preferred form of the vitamin for clinical use and is available as a colloidal suspension, an emulsion, and an aqueous suspension for parenteral use, and as a tablet for oral use. Menadione is available as a tablet for oral administration. A water-soluble form of menadione, menadiol sodium diphosphate, is also available for parenteral use. Vitamin K is widely used by the animal industry, particularly in poultry feeds. Because of the expense of phylloquinone, various water-soluble forms of menadione are the predominant sources used. Menadione forms a water-soluble sodium bisulfite addition product, menadione sodium bisulfite (MSB). In the presence of excess sodium bisulfite, MSB crystallizes as a complex with an additional mole of sodium bisulfite to form a compound called menadione sodium bisulfite complex (MSBC). This form has increased dietary stability compared to menadione and MSB and, therefore, is widely used in the poultry industry. A third water-soluble compound is a salt called menadione pyridinol bisulfite (MPB), which is formed by the addition of dimethylpyridinol. These three forms of menadione, and phylloquinone, have roughly equal biological activity on a molar basis in poultry diets. One of the menadione forms is usually added to the diets of laboratory animals and sometimes to those of swine.

ABSORPTION AND METABOLISM

Vitamin K is absorbed from the intestine into the lymphatic system of mammals by a process that requires the presence of both bile salts and pancreatic juice for optimal formation of mixed micellar structures. Shearer et al. (1974) have studied the absorption of radioactive phylloquinone. They found that fecal excretion of the unmetabolized form following a 1-mg dose of phylloquinone was less than 20 percent in normal subjects but increased to more than 70 percent in patients with impaired fat absorption. Animal diets could contain menadione, a mixture of menaquinones and phylloquinone, and there is evidence that the absorption of these various forms of vitamin K differs significantly. A series of investigations (Hollander, 1981) has demonstrated that phylloquinone is absorbed from the proximal small intestine by an energy-requiring process. On the other hand, menaquinone-9 is absorbed from the small intestine by a passive, noncarrier-mediated process. Mena-

FIGURE 8   Chemical structures of three major forms of vitamin K.

dione appears to be absorbed from both the colon and the small intestine by passive processes.

Suttie (1985a) has recently reviewed the distribution of various forms of vitamin K in tissues and the vitamin's metabolism. Almost 50 percent of a physiological dose of phylloquinone localizes in the liver at 3 hours after parenteral administration, while only 2 percent of a dose of menadiol diphosphate localizes in this tissue by that time. Although phylloquinone is rapidly concentrated in liver, it has a relatively short biological half-life. After a rapid drop during the first few hours following peak levels, injected phylloquinone has been shown to be removed from rat liver with a half-life of less than a day, suggesting very little long-term storage. Menaquinone-4 is the major lipophilic product of menadione metabolism formed when low doses of menadione are administered.

Menadione is rapidly metabolized, and three different urinary conjugates—the phosphate, sulfate, and glucuronide—of menadiol have been identified. The major route of excretion of intravenously administered radioactive phylloquinone appears to be fecal, with only a small percentage of fecal radioactivity present as unmetabolized phylloquinone. The metabolism of radioactive phylloquinone has now been studied in both the rat and in humans, and the glycones of the 5- and 7-carbon side chain carboxylic acid derivatives appear to be major excretion products. A significant amount of vitamin K stored in tissue is present as the 2,3-epoxide of the vitamin (Matschiner et al., 1970). It is likely that epoxide is also subjected to degraded metabolism before excretion. There are undoubtedly a number of metabolites and excretion products that have not yet been identified.

HYPERVITAMINOSIS

Early studies of vitamin K supplementation indicated the relative lack of toxic symptoms. Few systematic studies of the effects of excess vitamin administration have been carried out. Molitor and Robinson (1940) made a brief but rather comprehensive study of the toxicity of menadione and phylloquinone soon after the vitamin was discovered. Ansbacher et al. (1942), Smith et al. (1943), and Richards and Shapiro (1945) carried out standard pharmacological studies of menadione or menadione bisulfite in the next few years. These studies utilized standard laboratory animals and are summarized in Table 9. From these data it can be concluded that the LD_{50} for a single parenteral dose of menadione or its water-soluble derivative is in the range of 75 to 200 mg/kg of BW for chicks, mice, rats, rabbits, and dogs, and the LD_{50} for a single oral dose is 600 to 800 mg/kg of

BW at least for chicks and mice. These dosage levels are several orders of magnitude greater than the daily requirement of the vitamin. A very limited amount of data suggests that the chronic administration of a sublethal dose of menadione can produce hemolytic anemia. The only indication of an adverse response to vitamin K administration in domestic animals appears to be the recent report of Rebhun et al. (1984) of acute renal failure in horses following the parenteral administration of a single dose of menadione bisulfite. This response was reproduced experimentally, and the dosage used was within the range recommended by the manufacturer of the compound.

In contrast to the low but at least demonstrable toxicity of menadione, natural forms of vitamin K appear to be essentially innocuous. Molitor and Robinson (1940) administered up to 25 g/kg of BW of phylloquinone either parenterally or orally to laboratory animals with no reported adverse effect. Barash (1978) has reviewed reports of adverse effects on the cardiopulmonary system in humans following the intravenous injection of phylloquinone. This is an adverse drug reaction that occurs with very low frequency and may be a response to the colloid emulsion used as a vehicle rather than to the drug itself.

The toxicity of menadione is undoubtedly not related to its role as a precursor for tissue synthesis of an active form of vitamin K but because of its chemical properties as a quinone. Vitamin K is routinely administered to prevent hemorrhagic disease of the newborn. At one time, menadione was the form of the vitamin widely used. A high incidence of hemolytic anemia and liver toxicity following this therapy (Finkel, 1961; Barash, 1978) has led to the recommendation (American Academy of Pediatrics, 1971) of the administration of phylloquinone. The basis for the adverse reactions is not clear but is thought to be related to an influence on cellular redox state or sulfhydryl metabolism.

PRESUMED UPPER SAFE LEVELS

The limited amount of information available on vitamin K has failed to demonstrate any toxicity associated with the oral intake of phylloquinone and has shown that menadione can be ingested at levels as high as 1,000 times dietary requirements with no adverse effects.

SUMMARY

1. Phylloquinone, a natural form of vitamin K, exhibits no adverse effects when administered to animals in massive doses by any route.

TABLE 9 Research Findings of High Levels of Vitamin K in Animals

Species and No. of Animal	Age or Weight	Administration Amount/BW	Form	Duration	Route	Effect	Reference
Chicks, 55		Variable	Menadione	Single dose	Oral	LD_{50}, 804 mg/kg	Ansbacher et al., 1942
Chicks, 10	75 g	0.1 g/kg	Menadione	Single dose	IP	70% Mortality	Molitor and Robinson, 1940
Chicks, 10	75 g	0.15 g/kg	Menadione	Single dose	IP	90% Mortality	Molitor and Robinson, 1940
Chicks, 20	75 g	0.25 and 0.5 g/kg	Menadione	Single dose	IP	100% Mortality	Molitor and Robinson, 1940
Chicks, 20	75 g	25 g/kg	Phylloquinone	Single dose	IP	No effect	Molitor and Robinson, 1940
Dogs, 17		Variable	Menadione bisulfite	Single dose	IV	LD_{50}, 100–150 mg/kg	Richards and Shapiro, 1945
Dogs		15–40 mg/kg	Menadione bisulfite	Daily, 15 d	IV	Anemia; no mortality	Richards and Shapiro, 1945
Horses, 11	400 kg	2.1–8.3 mg/kg	Menadione bisulfite	Single dose	IM and IV	Acute renal failure	Rebhun et al., 1984
Mice, 60	20 g	0.4–0.8 g/kg	Menadione	Single dose	Oral	35–95% Mortality	Molitor and Robinson, 1940
Mice, 40	20 g	1.0 and 1.2 g/kg	Menadione	Single dose	Oral	100% Mortality	Molitor and Robinson, 1940
Mice, 64	20 g	Variable	Menadione	Single dose	Oral	LD_{50}, 620 mg/kg	Ansbacher et al., 1942
Mice, 20	20 g	50 mg/kg	Menadione	Single dose	IP	10% Mortality	Molitor and Robinson, 1940
Mice, 60	20 g	75, 100, and 150 mg/kg	Menadione	Single dose	IP	50–95% Mortality	Molitor and Robinson, 1940
Mice, 40	20 g	200–250 mg/kg	Menadione	Single dose	IP	100% Mortality	Molitor and Robinson, 1940
Mice, 26		Variable	Menadione bisulfite	Single dose	IV	LD_{50}, 250 mg/kg	Richards and Shapiro, 1945
Mice, 68	20 g	Variable	Menadione	Single dose	SC	LD_{50}, 138 mg/kg	Ansbacher et al., 1942
Mice, 60	20 g	15–25 g/kg	Phylloquinone	Single dose	Oral	No effect	Molitor and Robinson, 1940
Mice, 60	20 g	15–25 g/kg	Phylloquinone	Single dose	IP	No effect	Molitor and Robinson, 1940
Rabbits, 25		Variable	Menadione bisulfite	Single dose	IV	LD_{50}, 120 mg/kg	Richards and Shapiro, 1945
Rats, 12	Young	0.25 g/kg	Menadione	Daily, 30 d	Oral	No effect	Molitor and Robinson, 1940
Rats, 12	Young	0.35 g/kg	Menadione	Daily, 30 d	Oral	Anemia; no Mortality	Molitor and Robinson, 1940
Rats, 12	Young	0.5 g/kg	Menadione	Daily, 30 d	Oral	100% Mortality	Molitor and Robinson, 1940
Rats, 35	155–310 g	0.15–0.3 g/kg	Menadione bisulfite	Single dose	SC	LD_{50}, 175 mg/kg	Smith et al., 1943
Rats, 24	Young	0.35 and 2 g/kg	Phylloquinone	Daily, 30 d	Oral	No effect	Molitor and Robinson, 1940

2. The toxic level of menadione or its derivatives in the diet is at least 1,000 times the dietary requirement.

3. Menadione or its derivatives, when administered parenterally, have an LD_{50} in the range of a few hundred mg/kg of BW in all species studied except the horse. Doses of 2 to 8 mg/kg have been reported to be lethal in this species.

REFERENCES

Almquist, H. J. 1975. The early history of vitamin K. Am. J. Clin. Nutr. 28:656.

American Academy of Pediatrics, Committee on Nutrition. 1971. Vitamin K supplementation for infants receiving milk substitute infant formulas and for those with fat malabsorption. Pediatrics 48:483.

Ansbacher, S., W. C. Corwin, and B. G. H. Thomas. 1942. Toxicity of menadione, menadiol and esters. J. Pharmacol. Exp. Ther. 75:111.

Barash, P. G. 1978. Nutrient toxicities of vitamin K. Pp. 97–100 in Handbook in Nutrition and Food, M. Rechcigl, ed. New York: CRC Press.

Dam, H. 1929. Cholesterinstoffwechsel in Hühnereiern und Hüchnchen. Biochem. Z. 215:475.

Dam, H. 1964. The discovery of vitamin K, its biological functions and therapeutical application. Pp. 9–226 in Nobel Lectures Physiology or Medicine 1942–1962, Nobel Foundation, ed. New York: Elsevier.

Doisy, E. A., and J. T. Matschiner. 1970. Biochemistry of vitamin K. Pp. 293–331 in Fat Soluble Vitamins, R. A. Morton, ed. Oxford: Pergamon Press.

Esmon, C. T., J. A. Sadowski, and J. W. Suttie. 1975. A new carboxylation reaction. The vitamin K-dependent incorporation of $H^{14}CO_3^-$ into prothrombin. J. Biol. Chem. 25:4744.

Finkel, M. J. 1961. Vitamin K_1 and the vitamin K analogues. Clin. Pharmacol. Ther. 2:794.

Griminger, P. 1971. Nutritional requirements for vitamin K-animal studies. Pp. 39–59 in Symposium Proceedings on the Biochemistry, Assay, and Nutritional Value of Vitamin K and Related Compounds. Chicago: Association of Vitamin Chemists.

Hollander, D. 1981. Intestinal absorption of vitamins A, E, D, and K. J. Lab. Clin. Med. 97:449.

MacCorquodale, D. W., L. C. Cheney, S. B. Binkley, W. F. Holcomb, R. W. McKee, S. A. Thayer, and E. A. Doisy. 1939. The constitution and synthesis of vitamin K_1. J. Biol. Chem. 131:357.

Matschiner, J. T., R. G. Bell, J. M. Amelotti, and T. E. Knauer. 1970. Isolation and characterization of a new metabolite of phylloquinone in the rat. Biochim. Biophys. Acta 201:309.

Molitor, H., and J. Robinson. 1940. Oral and parenteral toxicity of vitamin K_1, phthiocol and 2-methyl 1,4-napthoquinone. Proc. Soc. Exp. Biol. Med. 43:125.

Nelsestuen, G. L., T. H. Zytkovicz, and J. B. Howard. 1974. The mode of action of vitamin K. Identification of γ-carboxyglutamic acid as a component of prothrombin. J. Biol. Chem. 249:6347.

Olson, R. E. 1980. Vitamin K. Pp. 170–180 in Modern Nutrition in Health and Disease, R. S. Goodhart and M. E. Shils, eds. Philadelphia: Lea & Febiger.

Rebhun, W. C., B. C. Tennant, S. G. Dill, and J. M. King. 1984. Vitamin K_3-induced renal toxicosis in the horse. J. Am. Vet. Med. Assoc. 184:1237.

Richards, R. K., and S. Shapiro. 1945. Experimental and clinical studies on the action of high doses of hykinone and other menadione derivatives. J. Pharmacol. Exp. Ther. 84:93.

Scott, M. L. 1966. Vitamin K in animal nutrition. Vit. Horm. 24:633.

Shearer, M. J., A. McBurney, and P. Barkhan. 1974. Studies on the absorption and metabolism of phylloquinone (vitamin K_1) in man. Vit. Horm. 32:513.

Smith, J. J., A. C. Ivy, and R. H. K. Foster. 1943. The pharmacology of two water-soluble vitamin K-like substances. J. Lab. Clin. Med. 28:1667.

Stenflo, J., P. Fernlund, W. Egan, and P. Roepstorff. 1974. Vitamin K dependent modifications of glutamic acid residues in prothrombin. Proc. Natl. Acad. Sci. U.S.A. 71:2730.

Suttie, J. W. 1985a. Vitamin K. Pp. 225–311 in The Fat-soluble Vitamins, A. T. Diplock, ed. London: William Heinemann Ltd.

Suttie, J. W. 1985b. Vitamin K-dependent carboxylase. Annu. Rev. Biochem. 54:459.

Ascorbic Acid

Ascorbic acid was recognized as early as 1734 as the factor in fresh fruit and vegetables that prevents the development of scurvy (Chick, 1953). Despite this early recognition, it was not until 1932 that two different research groups isolated and identified this compound from mammalian adrenal glands and citrus fruits (Svirbely and Szent-Gyorgyi, 1932a,b; Waugh and King, 1932). Ascorbic acid is a white crystalline compound, classified as a carbohydrate, with a molecular weight of 176 and a melting point of 190 to 192°C (Bauernfeind, 1982). It is readily soluble in water, slightly soluble in alcohol and glycerol, and virtually insoluble in ether and chloroform. Ascorbic acid is relatively stable in air. In aqueous solutions, however, ascorbic acid is attacked by oxygen and other oxidizing agents that convert the reduced form of the vitamin first to dehydroascorbic acid and then on to further oxidation products in irreversible reactions.

NUTRITIONAL ROLE

Dietary Requirements of Various Species

Ascorbic acid can be synthesized from glucose through the intermediate formation of glucuronic acid and gulonic acid (Burns, 1975). Ascorbic acid appears to be ubiquitous in all plants (Loewus et al., 1975). It is synthesized in all animal species studied with the exception of humans, several primates, the Indian fruit bat, the guinea pig, a few birds, fish, and invertebrates (Burns, 1957; Ray Chauduri and Chatterjee, 1969; Chatterjee et al., 1975). Therefore, ascorbic acid is not considered to be an essential dietary nutrient for most domestic animals and laboratory animals; however, it is physiologically essential for all of them.

Biochemical Functions

The basic functional property of ascorbic acid is its redox potential (+0.08 mV) whereby the compound can reduce transition metals, thus allowing the vitamin to participate in a number of metabolically important hydroxylation reactions. For example, a major function of ascorbic acid is as a cofactor in the biosynthesis of collagen. Research has indicated that the function of ascorbic acid in domestic animals is essentially the same as that in humans or guinea pigs. Furthermore, it would appear that there are certain circumstances or periods during which the biosynthesis of ascorbic acid in domestic animals may not be sufficient to meet metabolic demands. During such periods, such as those of disease or high environmental temperature, exogenous supplies of ascorbic acid have been shown to be beneficial to the health and survival of these animals (Cole et al., 1944; Rydell, 1948; Scott, 1975; Teare et al., 1979; Vaananen and Wekman, 1979; Bauernfeind, 1982).

FORMS OF THE VITAMIN

Vitamin C is available as ascorbic acid, ascorbate-2-sulphate, ascorbyl palmitate, and sodium ascorbate. However, for most animals requiring a dietary source of the vitamin, only ascorbic acid (see Figure 9 for reduced and oxidized forms) has significant antiscorbutic properties. For fish such as the catfish, salmon, and trout, there are reports that ascorbate-2-sulphate and ascorbyl palmitate also have antiscorbutic properties (Halver et al., 1975; Brandt et al., 1985).

ABSORPTION AND METABOLISM

The site and mechanism of absorption of ascorbic acid may differ between those animals capable of synthesiz-

```
        O                              O
        ‖                              ‖
        C ⌐                            C ⌐
        |  |                           |  |
  OH — C  |                      O = C  |
        ‖  |                           |  |
  OH — C  |                      O = C  |
        |  |                           |  |
   H — C — O                      H — C — O
        |                              |
  OH — C — H                    OH — C — H
        |                              |
        C  H₂OH                        C  H₂OH
```

| L-Ascorbic | L-Dehydroascorbic |
| acid | acid |

FIGURE 9   The reduced and oxidized forms of ascorbic acid.

ing the vitamin and those requiring it as dietary source (Hornig, 1975). In humans and guinea pigs, maximal absorption occurs in the duodenum (Nicholsen and Chornock, 1942; Hornig et al., 1973) by a Na^+-dependent, active, carrier-mediated system (Stevenson and Brush, 1969; Stevenson, 1974). In contrast, ascorbic acid absorption in the rat, for which it is not a dietary essential, is a passive process that occurs primarily in the ileum (Hornig et al., 1973). It is assumed that the absorption mechanism of ascorbic acid in domestic animals is similar to that of the rat (Spencer et al., 1963); however, there is very little information available on this subject. Research on ruminants (Knight et al., 1941; Cappa, 1958) and horses (Errington et al., 1954; Herrick, 1972; Jaeschke and Keller, 1982) indicates that the efficiency of ascorbic acid absorption by the oral route is low in these species due to ascorbic acid destruction by microbial action or other unknown factors.

The efficiency of enteric absorption of ascorbic acid in domestic animals has not been investigated. However, research on humans indicates that the efficiency of ascorbic acid absorption declines as the dosage levels are increased (Kubler and Gehler, 1970; Kallner et al., 1977; Hornig et al., 1980). More than 70 percent of intakes less than 1 g are absorbed, while intakes greater than 5 g have less than 20 percent absorption. In addition, recent research also indicates that the efficiency of absorption may decline with age (Davies et al., 1984). Women with an average age of 82.6 years appeared to have approximately one-tenth the absorptive capacity for ascorbic acid of women with an average age of 21.8 years.

Studies on the tissue distribution of ascorbic acid in domestic animals have not been extensive, with the exception of the chicken and the trout. In the trout, radioautographs of intubated trout indicated a heavy concentration of labeled ascorbic acid in the liver, the anterior kidney (which is also known as the head kidney or adrenal gland), the renal kidney, and, particularly, the skin and scales (Halver et al., 1975). Hilton et al. (1979) found the concentrations of ascorbic acid to be highest in the brain and gonads. In the chicken, the highest concentration of labeled ascorbic acid was observed in the liver with smaller amounts in the kidney, lung, and spleen (Hornig and Frigg, 1979). White muscle ascorbic acid concentrations were not affected by differences in the dietary level of ascorbic acid (Dorr and Nockels, 1971).

The metabolism of ascorbic acid in domestic animals has not been extensively studied. Presently, there is very little information available. The known metabolic fate of ascorbic acid varies with species and depends upon the route of introduction and the quantity taken in by the animal (Tolbert et al., 1975). In all species, initial metabolism involves the reversible conversion of ascorbic acid to dehydroascorbic acid; species' differences in the metabolism of ascorbic acid would appear to occur after this step. In the guinea pig and the rat, there is enzymatic delactonization of dehydroascorbic acid to diketogulonic acid (Kagawa et al., 1960, 1962), which can then be decarboxylated to form carbon dioxide and a number of other compounds (Hornig, 1975). Carbon dioxide exhalation is the major route of elimination of ascorbic acid by the guinea pig and the rat. In contrast, no enzymatic delactonization of dehydroascorbic acid has been found to occur in humans (Baker et al., 1966). Humans eliminate ascorbic acid and a number of its metabolites primarily by way of the urine (Hornig, 1975).

HYPERVITAMINOSIS

Despite the claims that ascorbic acid is nontoxic to humans (Pauling, 1970), a number of toxicity symptoms or signs in humans and a number of laboratory animals have been attributed to intakes of large doses of ascorbic acid. These include oxaluria (Keith et al., 1974), uricosuria (Stein et al., 1976), hypoglycemia (Lewin, 1974), excessive absorption of iron (Cook and Monsen, 1977), diarrhea, allergic responses, and increased activity of degradative enzymes of ascorbic acid (Schrauzer and Rhead, 1973), destruction of vitamin B_{12} (Herbert and Jacob, 1974), and interference with mixed-function oxidase systems in the liver (Peterson et al., 1982; Sutton et al., 1982). However, some of these abnormalities have been incidental and have been noted in uncontrolled experiments. There are a number of contradictory reports as well. For example, studies indicating the destruction of vitamin B_{12} by ascorbic acid (Herbert and

TABLE 10 Research Findings of High Levels of Ascorbic Acid in Animals

Species and No. of Animal	Age or Weight	Administration Amount	Duration	Route	Effect	Reference
Birds						
Chickens, 216	10 mo	0.4 g/kg diet	7 mo	Diet	No effect	Perek and Kendler, 1963
Chickens, 240	10 mo	0.4 g/kg diet	7 mo	Diet	No effect	Perek and Kendler, 1963
Chickens, 60	4 yr	1.2 g/kg diet	1 yr	Diet	No effect	Dorr and Nockels, 1971
Chickens, 24	2 yr	1.32 g/kg diet	>28 d	Diet	No effect	Chen and Nockels 1973
Chickens, 24	4 yr	1.32 g/kg diet	>28 d	Diet	No effect	Chen and Nockels, 1973
Chickens, 24	7 yr	1.32 g/kg diet	>28 d	Diet	No effect	Chen and Nockels, 1973
Chickens, 45	9 wk	1.33 g/kg diet	49 wk	Diet	No effect	Nockels, 1973
Chickens, 70	1 d	3.0 g/kg diet	32 wk	Diet	No effect	Schmeling and Nockels, 1978
Chickens, 24	30 wk	3.3 g/kg diet	>28 d	Diet	No effect	Chen and Nockels, 1973
Chickens, 200	1 d	3.3 g/kg diet	32 wk	Diet	No effect	Dorr and Nockels, 1971
Turkeys, 15–30		0.3 g/kg diet	47 d	Diet	No effect	Nestor et al., 1972
Dogs, 8		1 g/d	46 d	Oral	Aggravation of hypertrophic osteodystrophy	Teare et al., 1979
Dogs, 67		1–2.5 g/d	3–6 d	IV	No effect	Leveque, 1969
Dogs, 1	5 mo	0.5–3.0 g/d	14 d	IV	Improvement of osteodystrophy	Vaananen and Wekman, 1979
Dogs, 12		2 g/d	3–6 d	IV	No effect	Belfield, 1967
Dogs, 1	4 mo	3 g/d	12 d	IV	Improvement of osteodystrophy	Vaananen and Wekman, 1979
Dogs, 3		9 g/d	3 d	IV	No effect	Leveque, 1969
Fish						
Catfish, 60	25 g	5 g/kg diet	150 d	Diet	No effect	Mayer et al., 1978
Trout, 300	2 g	1 g/kg diet	16 wk	Diet	No effect	Hodson et al., 1980
Trout, 300	1 g	10 g/kg diet	16 wk	Diet	No effect	Lanno et al., 1985
Guinea pigs, 20	14 d (173 g)	8.7 g/kg diet	43 d	Diet	Decreased bone density and urinary hydroxyproline	Bray and Briggs, 1984

Species, no.	Weight/Age	Dose	Duration	Route	Effect	Reference
Guinea pigs, 6	350 g	0.2 g/d	112 d	Oral	Liver congestion	Osano and Myoga, 1981
Guinea pigs	250–300 g	0.3 g/d	4 d	Oral	Decreased MFO activity; altered phospholipid levels	Sutton et al., 1982
Guinea pigs, 5	Weanlings (<250 g)	0.15–0.5 mg/kg	70 d	Oral	Depressed 7-α-hydroxylase activity	Peterson et al., 1983
Horses, 2	390 and 533 kg	5 g/d	1 d	IM	No effect	Loscher et al., 1984
Horses, 4	440–600 kg	10 g/d	1 d	SC	No effect	Loscher et al., 1984
Horses, 20	300–400 kg	5–10 g/d	1 d	IV	No effect	Loscher et al., 1984
Horses, 10	276–609 kg	10–20 g/d	1 d	Drinking water	No effect	Loscher et al., 1984
Mice, 70	11 g	~0.012 g/d	4 wk	Drinking water	Decreased BW	Grunewald and Mitchell, 1981
Mink, 10		0.1–0.2 g/kg BW	>70 d	Diet	Anemia; reduced number and size of kits	Erdler and Helgebostad, 1972
Rats, 20	104 g	20 g/kg diet	10 mo	Diet	Increased urinary oxalate, liver and urinary iron	Keith et al., 1974
Rats, 7–10	Adult	1 g/kg	21 d	IV	Decreased body weight, serum thyroid hormone, and TSH	Marcusen and Heninger, 1976
Rats, 12	250–300 g	50 mg/d	21 d	SC	Alteration of sex organs and hormones	Paal and Duttagupta, 1978
Swine, 366	6 kg	0.9 g/kg diet	28 d	Diet	No effect	Mahan and Saif, 1983
Swine, 16	4–5 wk	0.99 g/kg diet	28 d	Diet	No effect	Yen and Pond, 1981
Swine, 62	21 d	1 g/d	10 d	Diet	No effect	Sandholm et al., 1979
Swine, 36		1 g/d	28 d	Drinking water	Increased growth	Brown et al., 1975
Swine, 26	11 kg	1 g/d	32 d	Drinking water	No effect	Brown et al., 1971
Swine, 12		1–10 g/kg	7 d	Diet	No effect	Chavez, 1983
Swine, 9	10–12 kg	1–8 g/d	6 wk	IM	No effect	Grondalin and Hansen, 1981

NOTE: In all cases, the form was ascorbic acid.

Jacob, 1974) have been questioned, and the potentiation of a vitamin B_{12} deficiency by high levels of ascorbic acid appears to be unlikely (Hogenkamp, 1980). Nevertheless, there seem to be some negative side effects to excessive amounts of ascorbic acid intakes, although there is no information indicating lethality. An LD_{50} value has not been determined for any laboratory animal species.

There is very little information on any toxicity signs associated with excess ascorbic acid intake in domestic animals. In studies conducted on dogs, Leveque (1969) reported allergic types of reaction in the mouth. These signs disappeared when the level of ascorbic acid intake by the dogs was reduced. However, this observation was incidental, and the study was not controlled. The only study conducted on mink suggested that these animals may be very sensitive to high levels of ascorbic acid (Ender and Helgebostad, 1972). Intakes of 100 to 200 mg of ascorbic acid/kg of BW per day produced a pronounced anemia in pregnant females with a subsequent significant reduction in the number and size of kits. Virtually no studies have been conducted on other species of domestic animals to determine the levels of ascorbic acid that may be toxic or the symptoms or signs of ascorbic acid toxicity.

Concentrations in Tissues

Ascorbic acid has a wide distribution in animal tissues. In laboratory animals, high concentrations of ascorbic acid are normally found in glandular tissues such as the pituitary, salivary, and adrenal glands with the adrenal glands having the greatest concentration (Kirk, 1962; Hammarstroem, 1966; Hornig, 1975). The brain, liver, lungs, pancreas, and spleen normally have intermediate to high concentrations. Because these organs are larger than glandular tissues, their contribution of ascorbic acid to the total amount in the body is far greater.

With the exception of fish, very few studies with domestic animal species have been concerned with the effect of excessive levels of ascorbic acid on tissue. In studies on catfish, salmon, and trout, the ascorbic acid concentrations in the liver and head kidney increased in relation to dietary levels of the vitamin, eventually showing a plateau beyond which there were no further increases (Halver et al., 1969; Hilton et al., 1978; Lim and Lovell, 1978). In contrast, muscle ascorbic acid concentrations were not affected by increases in the dietary intakes of the vitamin (Hilton et al., 1979). Similarly, studies on chickens have indicated that muscle ascorbic acid concentrations remained constant when the dietary level of ascorbic acid was varied (Dorr and Nockels, 1971). In virtually all other domestic animal studies, only plasma ascorbic acid levels have been reported.

They seldom have been related to differences in ascorbic acid intake levels.

PRESUMED UPPER SAFE LEVELS

As indicated in Table 10, there is insufficient information to determine maximum tolerance levels of ascorbic acid for most domestic animals species. At this time, the most complete information is available only for the chicken. Dietary intakes up to 3,300 mg of ascorbic acid/kg of feed do not appear to affect the chicken adversely in prolonged growth studies of more than 60 days (Herrick and Nockels, 1969; Chen and Nockels, 1973; Nockels, 1973; Schmeling and Nockels, 1978). Similarly, research on swine (Brown et al., 1971, 1975; Sandholm et al., 1979; Chavez, 1983) and fish (Lanno et al., 1985) has indicated that dietary intakes of as much as 10 g of ascorbic acid/kg of feed do not adversely affect growth. However, studies on swine have been of much shorter duration (less than 60 days) than studies on chickens or trout. In addition, intakes of 0.5 and 3.0 g ascorbic acid/day in cats and dogs, respectively, do not appear to affect these animals adversely in short-term studies (Belfield, 1967; Edwards, 1968; Leveque, 1969; Vaananen and Wekman, 1979). Long-term studies are required.

There is much more information concerning the safe and apparently toxic ascorbic acid intake levels in laboratory animal species (Table 10). Some of the results, such as those with rats and guinea pigs, are contradictory, however. Nevertheless, it appears that levels up to 1 g of ascorbic acid/kg of feed in rat and guinea pig diets do not adversely affect the animals' growth. Further studies are warranted.

SUMMARY

1. Excess ascorbic acid intakes in humans and laboratory animals have been reported to produce a variety of toxic signs or symptoms including allergic responses, oxaluria, uricosuria, and interference with mixed-function oxidase systems.

2. Despite the variety of toxic signs in laboratory animals due to excessive intakes of ascorbic acid, there has been no associated lethality reported.

3. There is insufficient information on the tolerance and toxicity of ascorbic acid in most domestic animals.

4. In growth studies of varying lengths, ascorbic acid intakes of 3.3 g/kg of feed for chickens and 10 g/kg of feed for swine and trout do not appear to affect adversely the growth of these animals. Studies on other domestic animals are needed.

REFERENCES

Baker, E. M., J. C. Saari, and B. M. Tolbert. 1966. Ascorbic acid metabolism in man. Am. J. Clin. Nutr. 19:371.

Bauernfeind, J. C. 1982. Ascorbic acid technology in agricultural, pharmaceutical, food and industrial applications. Pp. 395–497 in Ascorbic Acid: Chemistry, Metabolism and Uses, P. A. Seib and B. M. Tolbert, eds. Adv. Chem. Ser. Vol. 200. Washington, D.C.: American Chemical Society.

Belfield, W. O. 1967. Vitamin C in treatment of canine and feline distemper complex. Vet. Med. Small Anim. Clin. 62:345.

Brandt, T. M., C. W. Deyve, and P. A. Sieb. 1985. Alternate sources of vitamin C for channel catfish. Prog. Fish Cult. 47:55.

Bray, D. L., and G. M. Briggs. 1984. Decrease in bone density in young male guinea pigs fed high levels of ascorbic acid. J. Nutr. 114:920.

Brown, R. G., V. D. Sharma, L. G. Young, and J. G. Buchanan-Smith. 1971. Connective tissue metabolism in swine. II. Influence of energy level and ascorbate supplementation on hydroxyproline excretion. Can. J. Anim. Sci. 51:439.

Brown, R. G., J. G. Buchanan-Smith, and V. D. Sharma. 1975. Ascorbic acid metabolism in swine. The effects of frequency of feeding and level of supplementary ascorbic acid on swine fed various energy levels. Can. J. Anim. Sci. 55:353.

Burns, J. J. 1957. Missing step in man, monkey and guinea pig required for biosynthesis of L-ascorbic acid. Nature (London) 180:553.

Burns, J. J. 1975. Overview of ascorbic acid metabolism. Pp. 5–6 in Second Conference on Vitamin C. Ann. N.Y. Acad. Sci. 258.

Cappa, V. 1958. Destruction of vitamin C by the bacterial flora of the rumen. Riv. Zootec. 31:199.

Chatterjee, I. B., A. K. Majumber, B. K. Nandi, and N. Subramanian. 1975. Synthesis and some major functions of vitamin C in animals. Pp. 24–47 in Second Conference on Vitamin C. Ann. N.Y. Acad. Sci. 258.

Chavez, E. R. 1983. Supplemental value of ascorbic acid during late gestation on piglet survival and early growth. Can. J. Anim. Sci. 63:683.

Chen, A. A., and C. F. Nockels. 1973. The effect of dietary vitamin C, protein, strain and age on egg quality, production and serum and albumin protein of chickens. Poult. Sci. 52:1862.

Chick, H. 1953. Early investigations of scurvy and the anti-scorbutic vitamin. Proc. Nutr. Soc. 12:210.

Cole, C. L., R. A. Rasmussen, and F. Thorp, Jr. 1944. Dermatosis of the ears, cheeks, neck and shoulders of young calves. Vet. Med. 39:204.

Cook, J. D., and E. R. Monsen. 1977. Vitamin C, the common cold and iron absorption. Am. J. Clin. Nutr. 30:235.

Davies, H. E. F., J. E. W. Davies, R. E. Hughes, and E. Jones. 1984. Studies on the absorption of L-xyloascorbic acid (vitamin C) in young and elderly subjects. Hum. Nutr. Clin. Nutr. 38C:463.

Dorr, P. E., and C. F. Nockels. 1971. Effects of aging and dietary ascorbic acid on tissue ascorbic acid in the domestic hen. Poult. Sci. 50:1375.

Edwards, W. C. 1968. Ascorbic acid for treatment of feline rhinotracheitis. Vet. Med. Small Anim. Clin. Ther. 63:696.

Ender, F., and A. Helgebostad. 1972. Iron deficiency anemia in mink. Z. Tierphysiol. Tierernaehr. Futtermittelkd., Heft 19:22.

Errington, B. J., B. S. Hodgkiss, and E. P. Jayne. 1954. Ascorbic acid in certain body fluids of horses. Am. J. Vet. Res. 3:242.

Grondalin, T., and I. Hansen. 1981. Effect of mega doses of vitamin C on osteochondrosis in pigs. Nord. Veterinaermed. 33:423.

Grunewald, K. K., and L. K. Mitchell. 1981. Serum enzyme activities in mice fed a high level of ascorbic acid. Nutr. Res. 1:393.

Halver, J. E., L. M. Ashley, and R. E. Smith. 1969. Ascorbic acid

requirements of coho salmon and rainbow trout. Trans. Am. Fish Soc. 98:762.

Halver, J. E., R. R. Smith, B. M. Tolbert, and E. M. Baker. 1975. Utilization of ascorbic acid in fish. Pp. 81–102 in Second Conference on Vitamin C. Ann. N.Y. Acad. Sci. 258.

Hammarstroem, L. 1966. Autoradiographic studies on the distribution of C^{14}-labelled ascorbic acid and dehydroascorbic acid. Acta Physiol. Scand. 70 (Suppl. 289):1

Herbert, V., and E. Jacob. 1974. Destruction of vitamin B_{12} by ascorbic acid. J. Am. Med. Assoc. 230:241.

Herrick, J. B. 1972. Vitamin nutrition of the horse. Vet. Med. Small Anim. Clin. 67:688.

Herrick, J. B., and C. F. Nockels. 1969. Effect of a high level of dietary ascorbic acid on egg quality. Poult. Sci. 48:1518.

Hilton, J. W., C. Y. Cho, and S. J. Slinger. 1978. Effect of graded levels of supplemental ascorbic acid in practical diets fed to rainbow trout. J. Fish Res. Bd. Can. 35:431.

Hilton, J. W., R. G. Brown, C. Y. Cho, and S. J. Slinger. 1979. The synthesis, half-life and distribution of ascorbic acid in rainbow trout (*Salmo gairdneri*). Can. J. Fish Aquat. Sci. 37:170.

Hodson, P. V., J. W. Hilton, B. R. Blunt, and S. J. Slinger. 1980. Effects of dietary ascorbic acid on chronic lead toxicity to young rainbow trout (*Salmo gairdneri*). Can. J. Fish Aquat. Sci. 37:170.

Hogenkamp, H. P. C. 1980. The interaction between vitamin B_{12} and vitamin C. Am. J. Clin. Nutr. 33:1.

Hornig, D. 1975. Metabolism of ascorbic acid. World Rev. Nutr. Diet. 23:225.

Hornig, D., and M. Frigg. 1979. Effect of age on biosynthesis of ascorbic acid in chicks. Arch. Gefluegelkd. 43:108.

Hornig, D., F. Weber, and O. Wiss. 1973. Site of intestinal absorption of ascorbic acid in guinea pigs and rats. Biochem. Biophys. Res. Commun. 52:168.

Hornig, D., J. P. Vuilleumier, and D. Hartmann. 1980. Absorption of large, single, oral intakes of ascorbic acid. Int. J. Vit. Nutr. Res. 50:309.

Jaeschke, G., and H. Keller. 1982. The ascorbic acid status of horses. 4. Behaviour of intravenously applied ascorbic acid in the serum. Berl. Muench. Tieraerztl. Wochenschr. 95:71.

Kagawa, Y. 1962. Enzymatic studies on ascorbic acid catabolism in animals. I. Catabolism of 2,3-diketogulonic acid. J. Biochem. 51:134.

Kagawa, Y., Y. Mano, and N. Shimayino. 1960. Biodegradation of dehydro-2-ascorbic acid; 2,3-diketo-L-aldenic decarboxylase from rat liver. Biochim. Biophys. Acta 43:348.

Kallner, A., D. Hartmann, and D. Hornig. 1977. On the absorption of ascorbic acid in man. Int. J. Vit. Nutr. Res. 47:383.

Keith, M. O., B. G. Shah, E. A. Nera, and O. Pelletier. 1974. The effects of high ascorbic acid and iron intake on the renal excretion of oxalate, calcium, and iron and on the kidney of rats. Nutr. Rep. Int. 10:357.

Kirk, J. E. 1962. Variations in the tissue contents of vitamins and hormones. IV. Ascorbic acid. Vit. Horm. (N.Y.) 20:83.

Knight, C. A., R. A. Dutcher, N. B. Cuerrant, and S. I. Bechdel. 1941. Utilization and excretion of ascorbic acid in the dairy cow. J. Dairy Sci. 24:567.

Kubler, W., and J. Gehler. 1970. Für Kinetik der enteralen Ascorbin saure-Resorption zur Berechnung nicht dosisproportionaler Resorptionsrorgänge. Int. Z. Vit. Forschung. 40:442.

Lanno, R. P., J. W. Hilton, and S. J. Slinger. 1985. The effect of ascorbic acid on dietary copper toxicity in rainbow trout. Aquaculture 49:269.

Leveque, J. I. 1969. Ascorbic acid in treatment of the canine distemper complex. Vet. Med. Small Anim. Clin. 64:997.

Lewin, S. 1974. High intake of vitamin C in relation to adenosine

3':5'-cyclic monophosphate and guanosine 3':5'-cyclic monophosphate concentrations and to blood sugar concentrations. Biochem. Soc. Trans. 2:922.

Lim, C., and R. T. Lovell. 1978. Pathology of vitamin C deficiency syndrome in channel catfish. J. Nutr. 108:1137.

Loewus, F. A., G. Wagner, and J. C. Yang. 1975. Biosynthesis and metabolism of ascorbic acid in plants. Pp. 7–23 in Second Conference on Vitamin C. Ann. N.Y. Acad. Sci. 258.

Loscher, W., G. Jaeschke, and H. Keller. 1984. Pharmacokinetics of ascorbic acid in horses. Equine Vet. J. 16:59.

Mahan, D. C., and Z. J. Saif. 1983. Efficacy of vitamin C supplementation for weanling swine. J. Anim. Sci. 56:631.

Marcusen, D. C., and R. W. Heninger. 1976. Effect of ascorbic acid on the pituitary-thyroid system in the rat. J. Endocrinol. 70:313.

Mayer, F. L., P. M. Mehrle, and P. L. Crutcher. 1978. Interactions of toxaphene and vitamin C in channel catfish. Trans. Am. Fish. Soc. 107:326.

Nestor, K. E., S. P. Touchburn, and M. Treiber. 1972. The influence of dietary ascorbic acid on blood ascorbic acid level and egg production of turkeys. Poult. Sci. 51:1676.

Nicholsen, J. T. L., and F. W. Chornock. 1942. Intubation studies of the human small intestine. XXII. An improved technique for the study of absorption. Its application to ascorbic acid. J. Clin. Invest. 21:505.

Nockels, C. F. 1973. The influence of feeding ascorbic acid and sulfate on egg production and on cholesterol content of certain tissues of the hen. Poult. Sci. 53:373.

Ohno, T., and K. Myoga. 1981. The possible toxicity of vitamin C in guinea pigs. Nutr. Rep. Int. 24:291.

Paul, P. K., and P. N. Duttagupta. 1978. Beneficial or harmful effects of a large dose of vitamin C on the reproductive organs of the male rat depending upon the level of food intake. J. Exp. Biol. 16:18.

Pauling, L. 1970. Evolution and need for ascorbic acid. Proc. Natl. Acad. Sci. 67:1643.

Perek, M., and J. Kendler. 1963. Ascorbic acid as a dietary supplement for white leghorn hens under conditions of climatic stress. Br. Poult. Sci. 4:191.

Peterson, F. J., J. G. Babish, and J. M. Rivers. 1982. Excessive ascorbic acid consumption and drug metabolism in guinea pigs. Nutr. Rep. Int. 26:1037.

Peterson, F. J., D. E. Holloway, P. H. Duquette, and J. M. Rivers. 1983. Dietary ascorbic acid and hepatic mixed function oxidase activity in the guinea pig. Biochem. Pharm. 32:91.

Ray Chauduri, C., and I. B. Chatterjee. 1969. L-ascorbic acid synthesis in birds: Phylogenetic trend. Science 164:435.

Rydell, R. O. 1948. Dermatosis in calves. J. Am. Vet. Med. Assoc. 112:59.

Sandholm, M., T. Honkanen-Buyolski, and V. Rasi. 1979. Prevention of navel bleeding in piglets by preparturient administration of ascorbic acid. Vet. Rec. 104:337.

Schmeling, S. K., and C. F. Nockels. 1978. Effects of age, sex and ascorbic acid ingestion on chicken plasma corticosterone levels. Poult. Sci. 57:527.

Schrauzer, G. M., and W. J. Rhead. 1973. Ascorbic acid abuse: Effects of long-term ingestion of excessive amounts on blood levels and urinary excretion. Int. J. Vit. Nutr. Res. 43:201.

Scott, M. L. 1975. Environmental influences on the ascorbic acid requirements in animals. Pp. 151–155 in Second Conference on Vitamin C. Ann. N.Y. Acad. Sci. 258.

Sherlock, T. 1983. Tissue concentrations and proliferative effects of massive doses of ascorbic acid in the mouse. Nutr. Cancer 4:241.

Spencer, R. P., S. Purdy, R. Hoeldeke, T. M. Bow, and M. A. Markulus. 1963. Studies on intestinal absorption of L-ascorbic acid 1-^{14}C. Gastroenterology 44:768.

Stein, H. G., A. Hasan, and I. H. Fox. 1976. Ascorbic acid-induced uricosuria. A consequence of megavitamin therapy. Ann. Int. Med. 84:385.

Stevenson, N. R. 1974. Active transport of L-ascorbic acid in the human ileum. Gastroenterology 67:952.

Stevenson, N. R., and M. K. Brush. 1969. Existence and characteristics of Na$^+$-dependent active transport of ascorbic acid in guinea pigs. Am. J. Clin. Nutr. 22:318.

Sutton, J. L., T. K. Basu, and W. T. Dickerson. 1982. Effect of large doses of ascorbic acid on the mixed function oxidase system in guinea pig liver. Biochem. Pharmacol. 31:1591.

Svirbely, J. L., and A. Szent-Gyorgyi. 1932a. Hexuronic acid as the antiscorbutic factor. Nature (London) 129:576.

Svirbely, J. L., and A. Szent-Gyorgyi. 1932b. The chemical nature of vitamin C. Biochem. J. 26:865.

Teare, J. A., L. Krook, F. A. Kallfelz, and H. F. Hintz. 1979. Ascorbic acid deficiency and hypertrophic osteodystrophy in the dog: A rebuttal. Cornell Vet. 69:384.

Tolbert, B. M., M. Downing, R. W. Carlson, M. K. Knight, and E. M. Baker. 1975. Chemistry and metabolism of ascorbic acid and ascorbate sulfate. Pp. 48–69 in Second Conference on Vitamin C. Ann. N.Y. Acad. Sci. 258.

Vaananen, M., and L. Wekman. 1979. Scurvy as a cause of osteodystrophy. J. Small Anim. Pract. 20:491.

Waugh, N. A., and C. G. King. 1932. Isolation and identification of vitamin C. J. Biol. Chem. 97:325.

Yen, J. T., and W. G. Pond. 1981. Effect of vitamin C addition on performance, plasma vitamin C and hematic iron status in weanling pigs. J. Anim. Sci. 53:1292.

Thiamin

Jansen and Donath (1926) identified thiamin as the active factor in rice polishings and rice bran that prevents the disease beriberi in humans. In later studies, Kinnersly and Peters (1928) isolated thiamin from yeast and wheat germ. A synthetic process was developed for the production of this vitamin in 1936 (Robinson, 1966). Thiamin is a white hygroscopic crystalline compound that is stable at temperatures up to 100°C and is readily soluble in water. Thiamin consists of a pyrimidine nucleus and a thiazole ring linked by a methylene bridge (Williams and Cline, 1936) (see Figure 10).

NUTRITIONAL ROLE

Dietary Requirements of Various Species

Thiamin synthesis occurs only in plants and microbes; therefore, virtually all animals have nutritional requirements for this vitamin. There are a number of factors that can affect an animal's dietary requirement for the vitamin. Adult ruminants and horses can obtain adequate quantities of thiamin from bacteria in the rumen or cecum (Bechdel et al., 1926; McElroy and Gross, 1941; Kon and Porter, 1947; Hotzel and Barnes, 1966; Poe et al., 1972). However, young ruminants between the ages of 2 to 7 months can, under certain circumstances, develop polioencephalomalacia (cerebrocortical necrosis) in which thiamin deficiency plays an essential role (Edwin and Jackman, 1981/1982). This disease appears to result from either the ruminal destruction of formed thiamin or from the presence of antithiamin compounds there (Edwin and Jackman, 1981/1982).

Bacterial synthesis of thiamin in the cecum of nonruminant animals other than the horse can also occur. Rabbits and rats, which practice coprophagy, can obtain significant quantities of thiamin by this route. A deficiency state can be induced by feeding a thiamin-free or thiamin-inadequate diet, however (Wostmann et al., 1958; Reid et al., 1963; Bitter et al., 1969; Loew and Yert, 1976). Other factors such as age of the animal (Draper, 1958; Lazarov, 1977) and composition of the diet (Stirn et al., 1939; Wainio, 1942; Ellis and Madsen, 1944; Dicksen and Dahme, 1971; Holler et al., 1978) may also affect the thiamin requirement level of an animal. The thiamin requirements for most domestic animals, with the exception of horses and ruminants, range from 1 to 10 mg/kg of diet.

Biochemical Functions

The enzyme thiamin pyrophosphokinase and adenosine triphosphate (ATP) convert thiamin into its metabolically active coenzyme form, thiamin pyrophosphate (TPP) (Sauberlich, 1967). In the form of TPP, thiamin functions in the oxidative decarboxylation of α-keto acids, such as pyruvate and α-ketoglutarate. In addition, TPP functions in the transketolase reaction of the pentose phosphate pathway. Thiamin plays a very important role in glucose metabolism. Therefore, it is not surprising that the first signs of thiamin deficiency are usually of neurological origin. The thiamin status of an animal can also be determined by measurement of transketolase activity in erythrocytes or other tissues and of percentage of stimulation of that activity by exogenous TPP.

Thiamin also appears to be involved in nerve transmission and/or excitation, but whether this role involves TPP is not clear (Waldenlind, 1978). It is this function that appears to be related to the toxicity of thiamin (Itokawa, 1978).

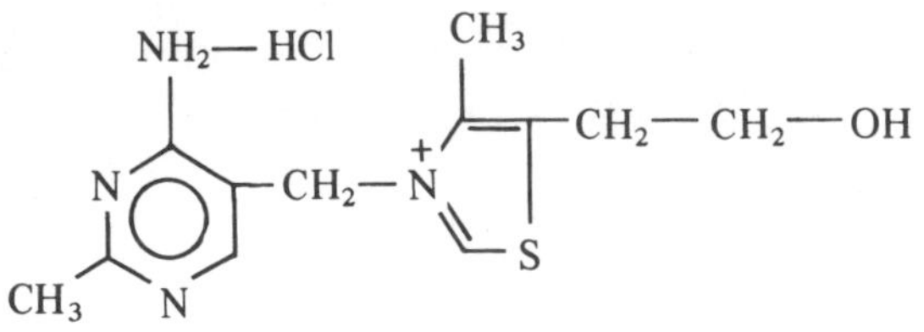

Thiamin Chloride-Hydrochloride

FIGURE 10   Chemical structure of thiamin hydrochloride.

FORMS OF THE VITAMIN

Thiamin is found in most animal tissues predominately in phosphorylated forms (e.g., thiamin mono-, di-, and triphosphates). In cereals and legumes it is present in a nonphosphorylated form. It is located predominately in the scutellum and germ of cereal grains and is, therefore, removed by milling. Thiamin hydrochloride and thiamin mononitrate are synthesized for commercial use in animal feeds. Thus, thiamin is found in the diet in one of three forms: free thiamin, phosphorylated thiamin, and protein-phosphate complexes.

ABSORPTION AND METABOLISM

In the gastrointestinal tract, the bound forms of thiamin are cleaved and the free form is absorbed primarily in the proximal small intestine (Sklan and Trostler, 1977). Thiamin absorption appears to be passive at high or pharmacological concentrations and active, by a carrier-mediated system, at low or physiological concentrations (Hoyumpa et al., 1975; Hoyumpa, 1982). Thiamin's absorption mechanism has only been studied in laboratory animals and humans; however, the absorption mechanism in domestic animals is assumed to be essentially the same.

The tissue distribution of thiamin appears to be fairly uniform, with higher concentrations found in the liver and kidney (Cohen et al., 1962; Hammarstrom et al., 1966; Ensminger et al., 1983a). Nervous tissues generally have low levels of thiamin. These tissues are able to conserve or maintain their thiamin more rigorously than other tissues, however (Robinson, 1966; Spector, 1982). In the human body, approximately 80 percent of the thiamin present is stored as TPP, 10 percent is stored as the triphosphate form, and the remainder is stored as the monophosphate form (Ensminger et al., 1983a). Thiamin is one of the most poorly stored of the vitamins. Mammals can exhaust their body stores within 1 to 2 weeks (Ensminger et al., 1983b). The pig, however, is an exception to this general rule. It can store large quantities of thiamin in skeletal muscles (Wilson et al., 1979). In mammals, excess thiamin is primarily eliminated by way of the urine in unaltered form (Kraus and Mahan, 1979). Nonetheless, a number of different metabolites of thiamin have been noted in rat and human urine (Neal and Pearson, 1964; Balaghi and Pearson, 1966; Sauberlich, 1967; Ariaey-Nejad et al., 1968).

HYPERVITAMINOSIS

The effects of excessive intakes of thiamin have been studied only in laboratory animals, dogs, and rabbits (Table 11). Lethal doses (i.e., LD_{100} levels) of thiamin by intravenous injection are 80 mg/kg of BW in mice, 180 mg/kg of BW in rabbits, and 50 to 125 mg/kg of BW in dogs (Haley and Flesher, 1946; Smith et al., 1947; Haley, 1948). Molitor (1942) has reported that lethal doses of thiamin in rats are 170 mg/kg of BW by intravenous administration and 9.5 g/kg of BW by oral administration. Furthermore, the LD_{50} values of thiamin in rats are 500 mg/kg of BW by subcutaneous administration and 6 g/kg of BW by oral administration (Molitor, 1942). The studies of Molitor are reported only in a review and consequently need to be confirmed. Nevertheless, it is obvious that the lethal effects of the vitamin are only produced at levels at least 1,000 times (by intravenous administration) that of the dietary requirement. In these studies, toxic levels of thiamin produced a wide variety of pharmacological effects in the animals. It appears that most of these toxic effects are produced only when the vitamin is administered acutely (Unna, 1972; Itokawa, 1978). There is virtually no information available to indicate any cumulative effects of thiamin administered chronically at levels below those that are acutely toxic.

In acute toxicity studies, excess thiamin appears to block nerve transmission, producing curare-like signs in the treated animal (Haley and Flesher, 1946; Smith et al., 1947, 1948; Haley, 1948; Hayashi et al., 1965). These general signs include restlessness, epileptiform convulsions, cyanosis, and labored respiration. Death from thiamin toxicity results from respiratory paralysis, usually accompanied by cardiac failure. In studies with dogs, artificial respiration applied after intravenous injection of a lethal dose of thiamin was partially effective in overcoming thiamin toxicity (Smith et al., 1948). The mechanism whereby high levels of thiamin block nerve transmission remains to be determined.

It should be noted that the majority of the studies conducted on the acute toxicity of parenterally administered thiamin have used the hydrochloride form of the vitamin. Therefore, it is possible that the observed toxicity signs may involve an altered acid-base balance due

TABLE 11 Research Findings of High Levels of Thiamin in Animals

Species and No. of Animal	Age or Weight	Amount (mg/kg of BW)	Reference
Dogs		50	Smith et al., 1948
Dogs, 12		~125	Smith et al., 1947
Mice, 150	22–42 g	80–92	Haley, 1948
Mice, 150	22–42 g	380–400	Haley, 1948
Rabbits, 5	? ? 1.7 kg	00 107	Haley, 1948
Rabbits, 5	1.5–1.9 kg	180–240	Haley and Flesher, 1946

NOTE: In all cases, the form was thiamin hydrochloride, the route was intravenous, a single dose was administered, and there was 100 percent mortality.

to the administration of excess chloride. This possibility requires further study.

The effect of varying levels of intake of thiamin on tissue levels of the vitamin has been studied only in the rat. In that species, thiamin concentrations in the brain, heart, and liver increased with thiamin intakes of as much as 1.5 to 2.0 mg/kg of BW/day, after which there were no further increases (Gubler and Murdock, 1982).

PRESUMED UPPER SAFE LEVELS

The maximum tolerable level of thiamin administered by either oral or parenteral routes has yet to be determined in most domestic animal species. In dogs, 100 to 115 μg of thiamin/kg of BW/day (oral route, Noel et al., 1977) and in sheep, 50 mg thiamin/head/day (intramuscular route, Yano and Kawashima, 1977) appeared safe. In rats, oral intakes of 50 to 100 mg/kg of diet appeared safe for periods up to 12 weeks (Morrison and Sarett, 1959; Schumacher et al., 1965; Itokawa and Fujiwara, 1973). Presently, it appears that for most species, dietary intakes of thiamin up to 1,000 times the requirement are apparently safe.

SUMMARY

1. There is little published information concerning thiamin tolerances and toxicity in domestic animal species.

2. Studies in laboratory animals, rabbits, and dogs indicate that parenteral administration of thiamin hydrochloride may be the only route by which signs of thiamin toxicity can be produced. In laboratory animals, rabbits, and dogs, thiamin toxicity is characterized by a depression of the respiratory center; however, the mechanism of this effect is unknown.

3. Acutely toxic levels of thiamin for laboratory ani-

mals given the vitamin parenterally range from approximately 80 to 400 mg/kg of BW.

4. Dietary intakes of thiamin up to 1,000 times the requirement level are apparently safe for most animal species.

REFERENCES

Ariaey-Nejad, M. R., and W. N. Pearson. 1968. 4-Methylthiazole-5-acetic acid urinary metabolite of thiamine. J. Nutr. 96:445.

Balaghi, M., and W. N. Pearson. 1966. Metabolism of physiological doses of thiazole-2-^{14}C-labeled thiamine by the rat. J. Nutr. 89:265.

Bechdel, S. I., C. H. Eckels, and L. S. Palmer. 1926. The vitamin B requirement of the calf. J. Dairy Sci. 9:409.

Bitter, R. A., C. J. Gubler, and R. W. Heninger. 1969. Effects of force-feeding on blood levels of pyruvate, glucocorticoids and glucose and on adrenal weight in thiamine-deprived and thiamine antagonist-treated rats. J. Nutr. 98:147.

Cohen, S., A. Uzan, and G. Valette. 1962. Thiamine et diathiopropylthiamine. Etude de leur metabolisme par marquage au soufre 35 chez la souris et le rat. Biochem. Pharmacol. 11:721.

Dicksen, G., and E. Dahme. 1971. Uber Klinik, Diagnose und Therapie der Cerebrocorticalnekrose (CCN) bei Kalb und Jungrind. Tieraerztl. Umsch. 26:517.

Draper, H. H. 1958. Physiological effects of aging. I. Efficiency of absorption and phosphorylation of radiothiamine. Proc. Soc. Exp. Biol. Med. 97:121.

Edwin, E. E., and R. Jackman. 1981/1982. Ruminant thiamine requirement in perspective. Vet. Res. Comm. 5:237.

Ellis, N. R., and L. L. Madsen. 1944. The thiamine requirement of pigs as related to the fat content of the diet. J. Nutr. 27:253.

Ensminger, A. H., M. E. Ensminger, J. E. Konlande, and J. R. K. Robson. 1983a. P. 2,415 in Foods and Nutrition Encyclopedia, Vol. 2, I-Z. California: Pegus Press.

Ensminger, A. H., M. E. Ensminger, J. E. Konlande, and J. R. K. Robson. 1983b. P. 1,208 in Foods and Nutrition Encyclopedia, Vol. 1, A-H. California: Pegus Press.

Gubler, C. J., and D. S. Murdock. 1982. Effect of treatment with thiamin antagonists, oxythiamin and pyrithiamin and of thiamin excess on the levels of distribution of thiamin in rat tissues. J. Nutr. Sci. Vitaminol. 28:217

Hammarstrom, L., H. Neujohr, and S. Ullberg. 1966. Autoradiographic studies on ^{35}S-thiamine distribution in mice. Acta Pharmacol. Toxicol. 24:24.

Haley, T. J. 1948. A comparison of the acute toxicity of two forms of thiamine. Proc. Soc. Exp. Biol. Med. 68:153.

Haley, T. J., and A. M. Flesher. 1946. A toxicity study of thiamine hydrochloride. Science 104:567.

Hayashi, T., Y. Kurahashi, and H. Takeuchi. 1965. Blocking action of thiamine and its derivatives upon neuromuscular transmission of cold-blooded animals. J. Vitaminol. 11:30.

Holler, H., E. L. Hindi, and G. Breeves. 1978. Einfuss von Thiamin (Vitamin B$_1$) auf mikrobielles Wochstum und Beldung von furchtigen Fettsauren in vitro in Pansensaft proteinfrei ernahter Schafe. Dtsch. Tieraerztl. Wochenschr. 85:200.

Hotzel, D., and R. H. Barnes. 1966. Contributions of the intestinal microflora to the nutrition of the host. Vitam. Horm. (N.Y.) 24:115.

Hoyumpa, A. M., Jr. 1982. Characterization of normal intestinal thiamin transport in man and animals. Pp. 337–343, in Thiamin: Twenty Years of Progress, Vol. 378, H. Z. Sable and C. J. Gubler, eds. New York: New York Academy of Sciences.

Hoyumpa, A. M., Jr., H. M. Middelton, F. A. Wilson, and S. Schenker. 1975. Thiamine transport across the rat intestine. I. Normal characteristics. Gastroenterology 68:1218.

Itokawa, Y. 1978. Effect of nutrient toxicities in animals and man: Thiamine. Pp. 3–23 in Nutrition and Food, Section E. Nutritional Disorders, Vol. 1, M. Rechcigl, Jr., ed. Florida: CRC Press Inc.

Itokawa, Y., and M. Fujiwara. 1973. Changes in tissue magnesium, calcium and phosphorus levels in magnesium-deficient rats in relation to thiamin excess or deficiency. J. Nutr. 103:438.

Jansen, B. C. P., and W. F. Donath. 1926. The isolation of anti-beriberi vitamin. Proc. K. Akad. Wetensch. Amsterdam 29:1390.

Kinnersly, H. W., and R. A. Peters. 1928. Antineuritic yeast concentrates. IV. The further purification of yeast vitamin B₁ (curative). Biochem. J. 22:419.

Kon, S. K., and J. W. G. Porter. 1947. The synthesis of vitamins in relation to requirements. Nutr. Abstr. Rev. 17:31.

Kraus, M. V., and L. K. Mahan, eds. 1979. Pp. 963 in Food, Nutrition and Diet Therapy, 6th ed. Toronto: W. B. Saunders and Co.

Lazarov, J. 1977. Resorption of vitamin B₁. XII. Changes in the resorption and phosphorylation of thiamine in rats in relation to age. Exp. Gerontol. 12:75.

Loew, F. M., and E. D. O. Yert. 1976. Thiamin status of laboratory rabbits. Lab. Anim. Sci. 26:201.

McElroy, L. W., and H. Gross. 1941. A quantitative study of vitamins in the rumen content of sheep and cows fed vitamin-low diets. J. Nutr. 21:163.

Molitor, H. 1942. Vitamins as pharmacological agents. Fed. Proc. 1:309.

Morrison, A. B., and H. P. Sarett. 1959. Effects of excess thiamine and pyridoxine on growth and reproduction in rats. J. Nutr. 69:111.

Neal, R. A., and W. N. Pearson. 1964. Studies on thiamine metabolism in the rat. I. Metabolic products found in the urine. J. Nutr. 83:343.

Noel, P. R. B., H. Chesterman, and D. W. Jolly. 1977. Thiamine (vitamin B₁) supplementation in the dog. Vet. Rec. 89:260.

Poe, S. E., G. E. Mitchell, and D. G. Ely. 1972. Rumen development in the lamb. III. Microbial B-vitamin synthesis. J. Anim. Sci. 34:826.

Reid, J. M., E. L. Hive, P. F. Braucher, and O. Mickelsen. 1963. Thiamine deficiency in rabbits. J. Nutr. 80:381.

Robinson, F. A., ed. 1966. P. 896 in Vitamin Co-Factors of Enzyme Systems. Toronto: Pergamon Press.

Sauberlich, H. E. 1967. Biochemical alterations in thiamine deficiency—their interpretation. Am. J. Clin. Nutr. 20:528.

Schumacher, M. F., M. A. Williams, and R. L. Lyman. 1965. Effects of high intakes of thiamine, riboflavin and pyridoxine on reproduction in rats and vitamin requirements of offspring. J. Nutr. 86:343.

Sklan, D., and N. Trostler. 1977. Site and extent of thiamine absorption in the rat. J. Nutr. 107:353.

Smith, J. A., P. P. Fao, and H. R. Weinstein. 1947. Some toxic effects of thiamine. Fed. Proc. 6:204.

Smith, J. A., P. P. Fao, and H. R. Weinstein. 1948. The curare-like action of thiamine. Science 108:412.

Spector, R. 1982. Thiamine homeostatis in the central nervous system. Pp. 344–354 in Thiamine: Twenty Years of Progress, Vol. 378. New York: New York Academy of Sciences.

Stirn, F. E., A. Arnold, and C. A. Elvehjem. 1939. The relation of dietary fat to the thiamin requirements of growing rats. J. Nutr. 17:485.

Unna, K. R. 1972. XII. Pharmacology and toxicology. Pp. 150–155 in The Vitamins, Vol. 5, W. H. Sebrell and R. S. Harris, eds. New York: Academic Press.

Wainio, W. W. 1942. The thiamine requirement of the albino rat as influenced by the substitution of protein for carbohydrate in the diet. J. Nutr. 24:317.

Waldenlind, L. 1978. Studies on thiamine and neuromuscular transmission. Acta Physiol. Scand. Suppl. 459:1.

Williams, R. R., and J. K. Cline. 1936. Synthesis of vitamin B. J. Am. Chem. Soc. 58:1504.

Wilson, E. D., J. H. Fisher, and P. A. Garcia, eds. 1979. Pp. 607 in Principles of Nutrition, 4th ed. New York: John Wiley & Sons.

Wostmann, B. L., P. L. Knight, and J. A. Reyniers. 1958. The influence of orally-administered penicillin on growth and liver thiamine of growing germfree and normal stock rats fed a thiamine-free diet. J. Nutr. 66:577–586.

Yano, H., and R. Kawashima. 1977. Effects of thiamine administration on blood lactic acid concentration and mineral metabolism in sheep. J. Nutr. Sci. Vitaminol. 23:491.

Niacin

The biochemical function of nicotinic acid was discovered before the nutritional role of this compound was appreciated. Warburg et al. (1935) isolated nicotinic acid from their "old yellow enzyme," subsequently identified as NADP (nicotinamide-adenine dinucleotide phosphate), and showed that it was part of a cellular hydrogen transport system (Warburg and Christian, 1936). Funk (1911) had previously isolated the compound in his search for the antipolyneuritis factor for the chick. After finding that nicotinic acid was not active in this animal model of beriberi, Funk dismissed it as being of little nutritional importance. It was not until Elvehjem et al. (1938) identified nicotinic acid as the factor that prevented "black tongue disease" in dogs that the nutritional role of the compound was recognized. Spies et al. (1938) soon demonstrated the importance of nicotinic acid in human health by showing that it cured pellagra.

NUTRITIONAL ROLE

Dietary Requirements of Various Species

Niacin is essential in the diets of nonruminant species for the prevention of a variety of severe metabolic disorders of the skin, gastrointestinal tract, and other organs. The first signs of niacin deficiency in most species are loss of appetite, reduced growth, generalized muscular weakness, digestive disorders, and diarrhea. A scaly dermatitis and, often, a microcytic anemia follow these signs. These conditions are referred to as black tongue disease in dogs, pellagra in humans, and pig pellagra in swine. The niacin-deficient chick also shows an abnormality of leg development called perosis. The niacin requirements of animals range from about 11 mg/kg of diet for dogs to 45 mg/kg of diet for cats. A primary

determinant of this variation is in the efficiency of metabolic conversion of tryptophan to niacin. Ruminants are usually capable of deriving all of their required niacin from ruminal microbial synthesis. Microbial synthesis is via the quinolinic acid pathway as well as from tryptophan.

Biochemical Functions

The biochemical bases for the diverse effects of niacin deficiency involve the numerous metabolic reactions, which in turn involve nicotinamide. These include some 35 oxidation-reduction reactions in which nicotinamide participates as either of the pyridine nucleotides (NAD[H] or NADP[H]) acting as two-electron transporters. NADH transfers electrons from metabolic intermediates to the mitochondrial electron transport chain, while NADH and NADPH serve as reducing agents in a large number of biosynthetic processes. Thus, nicotinamide has physiologically critical roles in mitochondrial respiration and in the metabolism of carbohydrates, lipids, and amino acids.

FORMS OF THE VITAMIN

Niacin is the accepted term used as the generic descriptor of pyridine 3-carboxylic acids and their derivatives that exhibit the biological activity of the amide of nicotinic acid—in other words, nicotinamide. Of the compounds with niacin activity, nicotinic acid and nicotinamide show the greatest biological potency (see Figure 11). Some analogs such as 3-acetyl pyridine and pyridine β-sulfonic acid show niacin-antagonistic activities. Niacin is widely distributed in foods of either plant or animal origin. Cereals comprise the most important sources of niacin in most animal diets. Much of that niacin appears to be present in bound forms with limited

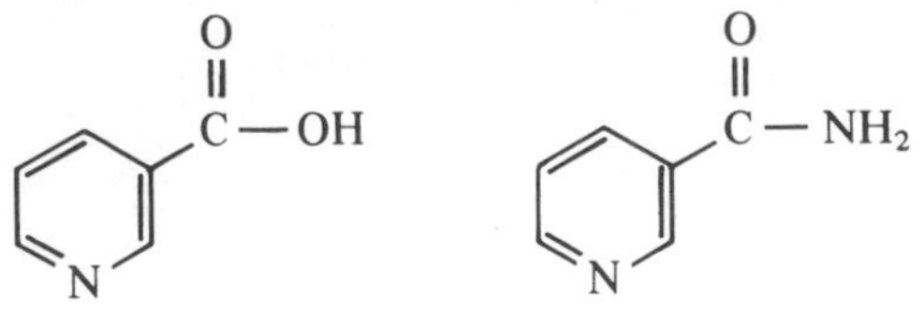

Nicotinic acid   Nicotinamide

FIGURE 11 Chemical structures of major niacin-active compounds.

availability to animals, however. Ghosh et al. (1963) found that most of the niacin in cereals and about 40 percent in oilseeds is bound. In contrast, meats, fish, and milk contain no bound forms of niacin. Bound niacin is released, however, by alkaline treatment of food materials.

Many cereals and other feedstuffs also contain relative excesses of the amino acid leucine, an antagonist of the metabolic conversion of tryptophan to niacin. For these reasons, niacin is generally added as nicotinic acid or nicotinamide to mixed feeds for nonruminants to ensure nutritional adequacy. Although it appears that ruminant species are able to obtain sufficient amounts of niacin from their rumen microflora, recent studies have also indicated some benefits of niacin supplementation in ruminant feeds in some circumstances (Frank and Schultz, 1979; Gulbert and Huber, 1979; Riddell et al., 1981; Jaster et al., 1983).

ABSORPTION AND METABOLISM

Nicotinic acid and nicotinamide are absorbed almost completely by simple diffusion across the intestinal mucosa. The rate of diffusion of nicotinic acid is about half that of nicotinamide. Absorbed nicotinic acid is thought to be converted to the amide form in the intestinal mucosa. Nicotinamide is taken up by tissues and incorporated into its coenzyme forms of NADH and NADPH. This process has been found to be regulated in neurons such that the cellular coenzyme levels are controlled by the uptake of niacin from the extracellular fluid.

Many of the pyridine nucleotides of tissues are synthesized from niacin derived from the metabolism of the amino acid tryptophan. Because tryptophan can be converted to nicotinamide mononucleotide (NMN) and hence to NAD, the efficiency of this conversion is a primary determinant of the dietary niacin requirement of specific animals. Tryptophan-niacin conversion tends to be high at low levels of intake of tryptophan and decreases with increasing levels of intake. Other factors including the intake of leucine, total protein, and pyridoxine also affect this conversion. In humans, the conversion is such that about 60 mg of dietary tryptophan is equivalent to about 1 mg of niacin. The efficiency of this conversion varies considerably among species according to the activity of picolinic acid carboxylase. This enzyme degrades an intermediate in the pathway (α-amino-β-muconic-ϵ-semialdehyde), thus reducing the yield of NMN. Species such as the cat have relatively high picolinic carboxylase activities. Cats also have high dietary requirements for niacin because it is difficult for them to produce the vitamin metabolically from tryptophan.

Niacin metabolites are readily excreted in the urine in proportion to the immediate level of vitamin intake. At low doses, the primary excretory forms are nicotinic acid and nicotinamide. At higher doses, however, several metabolites are excreted in the urine. The particular pattern of metabolites varies according to species. For example in rats the metabolite pattern is N_1-methylnicotinamide, nicotinuric acid, and nicotinamide-N_1-oxide; in humans, N_1-methylnicotinamide, N_1-methyl-2-pyridone-carboxamide, N_1-methyl-4-pyridone-3-carboxamide, and nicotinamide-N_2-oxide. Little niacin is retained in the body; most is excreted in the urine in a short time. In the case of high doses of niacin, 75 to 90 percent of the dose is usually excreted within 24 hours.

HYPERVITAMINOSIS

High levels (Table 12) of nicotinic acid, such as 3 g/day in humans, can cause vasodilation, itching, sensations of heat, nausea, vomiting, headaches, and occasional skin lesions (Robie, 1967; Hawkins, 1968). Hankes (1984) stated that oral doses of nicotinic acid as great as 100 g have been given without causing more severe reactions; however, the case report of Winter and Boyer (1973) indicates that hepatotoxicity can be produced in humans at dosages of 3 to 9 g of nicotinamide/day. Winter and Boyer found that doses of nicotinamide of about 9 g/day caused nausea and vomiting followed by elevations in serum transaminases, alkaline phosphatase, and total bilirubin. These signs were associated with portal fibrosis of the liver in an individual that had taken 9 g of nicotinamide/day for several months. Upon discontinuing the high level of nicotinamide intake, clinical parameters of liver function returned to normal within 22 days.

A threshold of 1 to 3 mg of nicotinic acid/kg of BW in the guinea pig produces vasodilation, which is shown by a cutaneous flush (Andersson et al., 1977). This effect is associated with elevation in skin temperature in the ears and increases in cyclic adenosine 3',5'-monophosphate (AMP) levels in that tissue. The threshold dose for flushing in humans is probably around 250 mg of nicotinic acid (Horrobin, 1980). The stimulation in production of a prostaglandin may produce the skin flush, which is indicated by the findings that the reaction in humans is reduced by pretreatment with indomethacin, an inhibitor of prostaglandin synthetase. In addition, flushing also can be produced in humans by administration of either cyclic-AMP or prostaglandin E_1 (Andersson et al., 1977; Svedmyr et al., 1977).

High intake levels (more than 3 g/day) of nicotinic acid have been shown to affect serum cholesterol and lipoprotein levels in humans. Although such treatment reduces levels of very-low-density (VLDL), intermediate-density (IDL), and low-density (LDL) lipoproteins, it increases levels of high-density (HDL) lipoproteins. The basis of the latter effect appears to be reduction of HDL catabolism (Blum et al., 1977). For this reason, nicotinic acid has been used in the treatment of hyperlipidemias (Patsch et al., 1977; Smith, 1981). Side effects of high levels of treatment, such as 300 mg of nicotinic acid/kg of BW/day for 3 weeks, have been observed in the normocholesterolemic rat. These include rebounds in plasma-free fatty acids and triglycerides, and triglyceride accumulation in liver (Subissi et al., 1980). Although the hypolipidemic effect of nicotinamide appears to be much less than that of nicotinic acid, studies in rats (Dalton et al., 1970) have shown that because of the much longer serum half-life of nicotinamide, its hypolipidemic effect is much longer.

The intake of 1 g or more of nicotinic acid has been found to reduce the urinary clearance of uric acid (Gershon and Fox, 1974). This effect is thought to be involved in the hyperuricemia frequently observed during the administration of 3 g/day of nicotinic acid for treatment of schizophrenia in humans (Hankes, 1984).

Studies on experimental animals have shown that the animal's exposure to high levels of nicotinamide can affect the metabolism of xenobiotic agents. Kamat et al. (1980) showed that the intraperitoneal administration of 100 mg nicotinamide/kg of BW in rats was effective in inducing the hepatic microsomal mixed function oxygenase (MFO) system (namely, NADPH-cytochrome c reductase, cytochrome P-450, and cytochrome b_5), and several drug-metabolizing enzyme systems (including aryl hydrocarbon hydroxylase, aminopyrine N-demethylase, and uridine 5'-diphosphate (UDP) glucuronosyl transferase). It is likely that the following may relate to the altered metabolism of the active agents by the effect of nicotinamide on the MFO system: potentiation of anti-epileptic activity of phenobarbital (Bourgeois et al., 1983); prevention of organophosphate-induced micromelia in the embryonic chick (Byrne and Kitos, 1983); protection from some acute effects of certain hepatocarcinogens (Schoental, 1977);

reduction of tumorigenesis induced by bracken fern (Pamukuo et al., 1981) or diethylnitrosamine (Schoental, 1977); and protection from pancreatic islet cell damage due to the diabetigenic substance streptozotocin (Wilander and Gunnarsson, 1975; Wick et al., 1977; Kazumi et al., 1978; Yoshino et al., 1979). Most of these effects have been observed in animals treated with nicotinamide at levels of 250 to 500 mg/kg of BW or fed the vitamin at 0.5 percent of the diet.

Chen et al. (1938) reported the toxicity of nicotinic acid for dogs. They found that repeated oral administration of 2 g/day of nicotinic acid (133 to 145 mg/kg of BW) produced bloody feces in a few dogs. Convulsions and death followed. Doses of nicotinic acid as great as 0.5 g/day, which is about 36 mg/kg of BW, produced slight proteinuria after 8 weeks. Hoffer (1969) has presented the median lethal doses of nicotinamide in g/kg of BW for several species. For the mouse, the median lethal doses are 4.5 to 7 g orally, 2.5 to 4.5 g intravenously, and 2.8 g by subcutaneous injection; for the rat, 5 to 7 g orally and 4 to 5 g intravenously; and for the rabbit, 2.5 g intravenously. However, there have been very few animal studies upon which to base estimates of the toxicity of high doses of niacin. Studies by Toth (1983) indicated that life-long exposures of mice to high levels of nicotinamide were not carcinogenic. Baker et al. (1976) showed that dietary levels of nicotinamide above 5,000 mg/kg depressed the growth of chicks, but that dietary levels of nicotinic acid as great as 20,000 mg/kg did not affect growth. Certain derivatives of niacin, such as 6-aminonicotinamide, isonicotinic acid, and isonicotinic hydrazide, have been shown to be lethal, teratogenic, and/or carcinogenic (Matschke and Fagerstone, 1977; Tsarichenko et al., 1977; Zackheim, 1978; Uyeki et al., 1982; Toth, 1983). The local toxicity of 6-aminonicotinamide is the basis of its therapeutic use for psoriasis (Zackheim, 1978).

PRESUMED UPPER SAFE LEVELS

Estimates of maximum tolerable levels of niacin-active compounds are not possible because of the limited definitive quantitative data presently available. That evidence suggests that levels greater than approximately 350 to 500 mg of nicotinic acid equivalents/kg of BW/day may be toxic. Because nicotinic acid is well absorbed, limits of safe exposure of niacin-active compounds are expected to be similar for oral and parenteral administration. The level of 350 mg nicotinamide/kg of BW/day is presumed safe for chronic exposure. Nicotinic acid may be tolerated at intakes as great as four times this level.

TABLE 12 Research Findings of High Levels of Niacin in Animals

Species and No. of Animal	Age or Weight	Administration Amount	Form	Duration	Route	Effect	Reference
Chickens	3–17 d of duration	Graded levels	6-Amino-nicotinamide		Daily injections into yolk sac	LD$_{50}$, 0.073 µmole/egg; growth retardation; anteriorly directed, short legs; coarse, dense feathering	Uyeki et al., 1982
Chickens	3–17 d of duration	Graded levels	6-Diethylamino-nicotinamide		Daily injections into yolk sac	LD$_{50}$, 0.23 µmole/egg; growth retardation; anteriorly directed, short legs; coarse, dense feathering	Uyeki et al., 1982
Chickens	3–17 d of duration	Graded levels	6-Methylamino-3-(N-methyl)-nicotinamide		Daily injections into yolk sac	LD$_{50}$, 1.11 µmole/egg; growth retardation; anteriorly directed, short legs; coarse, dense feathering	Uyeki et al., 1982
Chickens	3–17 d of duration	Graded levels	6-Dimethylaminoni-cotinamide		Daily injections into yolk sac	LD$_{50}$, 1.32 µmole/egg; growth retardation; anteriorly directed, short legs; coarse, dense feathering	Uyeki et al., 1982
Chickens	3–17 d of duration	Graded levels	6-Chloro-3[N-(5-diethylamino)-2-pentyl]-nicotinamide		Daily injections into yolk sac	LD$_{50}$, 3.00 µmole/egg; growth retardation; anteriorly directed, short legs; coarse, dense feathering	Uyeki et al., 1982
Chickens, 6/group	87 g, average initial weight	5,000–20,000 mg/kg diet	Nicotinamide	8–16 d, posthatching	Diet	Depressed weight gain linearly over range	Baker et al., 1976
Humans, 99		1% in gel	6-Amino-nicotinamide	4 wk	Topical	Transient tachyphylaxis observed in 10 subjects; slight mucocutaneous toxicity observed in 25 subjects	Zackheim, 1978

51

Species	Dose	Age	Compound	Duration	Route	Effect	Reference
Mice	Graded doses		Nicotinamide	Single dose	IV	LD$_{50}$, 2.5–4.5 g/kg BW	Hoffer, 1969
Mice	Graded doses		Nicotinamide	Single dose	SC	LD$_{50}$, 2.8 g/kg BW	Hoffer, 1969
Mice	Graded doses		Nicotinamide	Single dose	Oral	LD$_{50}$, 4.5–7.0 g/kg BW	Hoffer, 1969
Mice, dams, 20	Graded doses		6-Amino-nicotinamide	8–10 d, postpartum	Oral	LD$_{50}$, 19 mg/kg at 8 d; increased to 100 mg/kg by 13 d; visceral and gross anomalies (hydrocephalus, cleft palate, short jaws) increased with dose	Matschke and Fagerstone, 1977
Mice	15–30 mg/m^3		Isonicotinic acid	Single dose	IP	Stimulated nervous system; increased liver and myocardial dystrophy; lymphostatis in lymph vessels	Tsarichenko et al., 1977
Mice	Graded doses		Isonicotinic acid	Single dose	IP	LD$_{50}$, 436.2 mg/kg BW	Tsarichenko et al., 1977
Mice	Graded doses		Isonicotinic acid	Single dose	Oral	LD$_{50}$, 3,122.8 mg/kg BW	Tsarichenko et al., 1977
Mice, females, 50 males, 50	1% in drinking water	6 wk	Nicotinamide	Lifespan	Oral	No effects on longevity or tumor incidence	Toth, 1983
Mice, females, 50 males, 50	1% in drinking water	6 wk	Isonicotinamide	Lifespan	Oral	No effects on longevity or tumor incidence	Toth, 1983
Rabbits	Graded doses		Nicotinamide	Single dose	IV	LD$_{50}$, 2.5 g/kg BW	Hoffer, 1969
Rats	Graded doses		Nicotinamide	Single dose	IV	LD$_{50}$, 4.0–5.0 g/kg BW	Hoffer, 1969
Rats	Graded doses		Nicotinamide	Single dose	Oral	LD$_{50}$, 5.0–7.0 g/kg BW	Hoffer, 1969

SUMMARY

1. Niacin is the generic description for compounds required by all animals for the metabolic production of essential metabolic electron carriers NAD(H) and NADP(H).

2. Limited research indicates that nicotinic acid and nicotinamide are toxic at dietary intakes greater than about 350 mg/kg of BW/day.

REFERENCES

Andersson, R. G. G., G. Aberg, R. Brattsand, E. Ericsson, and L. Lundholm. 1977. Studies on the mechanism of flush induced by nicotinic acid. Acta Pharmacol. Toxicol. 41:1.

Baker, D. H., J. T. Yen, A. H. Jensen, R. G. Teeter, E. N. Michel, and J. H. Burns. 1976. Niacin activity in niacinamide and coffee. Nutr. Rep. Int. 14:115.

Blum, C. B., R. I. Levy, S. Eisenberg, M. Hall, R. H. Goebel, and M. Berman. 1977. High density lipoprotein metabolism in man. J. Clin. Invest. 60:795.

Bourgeois, B. F. D., W. E. Dobson, and J. A. Ferrendelli. 1983. Potentiation of the antiepileptic activity of phenobarbital by nicotinamide. Epilepsia 24:238.

Byrne, D. H., and P. A. Kitos. 1983. Teratogenic effects of cholinergic insecticides in chick embryos. Biochem. Pharmacol. 32:2881.

Chen, K. K., C. L. Rose, and E. B. Robbins. 1938. Toxicity of nicotinic acid. Proc. Soc. Exp. Biol. Med. 38:241.

Dalton, C., T. C. Van Trabert, and J. X. Dwyer. 1970. Relation of nicotinamide and nicotinic acid to hypolipidemia. Biochem. Pharmacol. 19:2609.

Elvehjem, C. A., R. J. Madden, F. M. Strong, and D. W. Woopley. 1938. The isolation and identification of the anti-black tongue factors. J. Biol. Chem. 123:137.

Frank, T. J., and L. H. Schultz. 1979. Oral nicotinic acids as a treatment for ketosis. J. Dairy Sci. 62:1804.

Funk, C. 1911. On the chemical nature of the substances which cure polyneuritis in birds induced by a polished rice diet. J. Physiol. 43:395.

Gershon, S. L., and I. H. Fox. 1974. Pharmacologic effects of nicotinic acid on human purine metabolism. J. Lab. Clin. Med. 84:179.

Ghosh, H. P., P. K. Sarkar, and B. C. Guha. 1963. Distribution of the bound form of nicotinic acid in natural materials. J. Nutr. 79:451.

Gulbert, K., and J. T. Huber. 1979. Influence of supplemental niacin with and without non-protein nitrogen in the performance of lactating dairy cows. J. Dairy Sci. 62:78.

Hankes, L. V. 1984. Nicotinic acid and nicotinamide. Pp. 329–377 in Handbook of Vitamins, Nutritional, Biochemical, and Clinical Aspects. L. J. Machlin, ed. New York: Marcel Dekker.

Hawkins, D. R. 1968. Treatment of schizophrenia based on the medical model. J. Schiz. 2:3.

Hoffer, A. 1969. Safety, side effects and relative lack of toxicity of nicotinic acid and nicotinamide. Schizophrenia 1:78.

Horrobin, D. F. 1980. Schizophrenia: A biochemical disorder? Biomedicine 32:54.

Jaster, E. H., G. F. Hartwell, and M. F. Hutjens. 1983. Feeding supplemental niacin for milk production in six dairy herds. J. Dairy Sci. 66:1046.

Kamat, J. P., L. M. Narurkar, N. A. Mhatre, and M. V. Narurkar. 1980.

Nicotinamide induced hepatic microsomal mixed function oxidase system in rats. Biochem. Biophys. Acta 628:26.

Kazumi, T., G. Yoshino, S. Fuji, and S. Baba. 1978. Tumorigenic action of streptozotocin on the pancreas and kidney in male Wistar rats. Cancer Res. 38:2144.

Matschke, G. H., and K. A. Fagerstone. 1977. Teratogenic effects of 6-aminonicotinamide in mice. J. Toxicol. Environ. Health 3:735.

Pamukuo, A. M., U. Milli, and G. T. Bryan. 1981. Protective effect of nicotinamide on bracken fern induced carcinogenicity in rats. Nutr. Cancer 3:86.

Patsch, J. R., D. Yeshurm, R. L. Jackson, and A. M. Gotto. 1977. Effects of clofibrate, nicotinic acid and diet on the properties of the plasma lipoproteins in a subject with type III hyperlipoproteinemia. Am. J. Med. 63:1001.

Riddell, D. O., E. E. Bartley, and A. D. Dayton. 1981. Effect of nicotinic acid on microbial protein synthesis in vitro and on dairy cattle and milk production. J. Dairy Sci. 65:782.

Robie, T. R. 1967. Cyproheptadine: An excellent antidote for niacin-induced hyperthermia. J. Schiz. 1:133.

Schoental, R. 1977. The role of nicotinamide and of certain other modifying factors in diethylnitrosamine carcinogenesis: Fusaria mycotoxins and "spontaneous" tumors in animals and man. Cancer 40:1833.

Smith, S. R. 1981. Severe hypertriglyceridemia responding to insulin and nicotinic acid therapy. Postgrad. Med. J. 57:511.

Spies, T. D., C. Copper, and M. A. Blankenhorn. 1938. The use of nicotinic acid in the treatment of pellagra. J. Am. Med. Assoc. 110:622.

Subissi, A., P. Schiantrelli, M. Biagi, and G. Sardelli. 1980. Comparative evaluation of some pharmacological properties and side effects of D-glucitol hexa nicotinate sorbinicate and nicotinic acid correlated with plasma concentration of nicotinic acid. Atherosclerosis 36:135.

Svedmyr, M., A. Heggelund, and G. Aberg. 1977. Influence of indomethacin on flush induced by nicotinic acid in man. Acta Pharmacol. Toxicol. 41:397.

Tsarichenko, G. V., V. I. Bobrov, and M. V. Starkov. 1977. Toxicity of isonicotinic acid. Khom-Farm Zh. 11:45.

Toth, B. 1983. Lack of carcinogenicity of nicotinamide and isonicotinamide following lifelong administration to mice. Oncology 40:72.

Uyeki, E. M., J. Doull, C. C. Cheng, and M. Misawa. 1982. Teratogenic and antiteratogenic effects of nicotinamide derivatives in chick embryos. J. Toxicol. Environ. Health 9:963.

Warburg, O., and W. Christian. 1936. Pyridine, the hydrogen-transferring component of the fermentation enzymes (pyridine nucleotide). Biochem. Z. 287:291.

Warburg, O., W. Christian, and A. Griese. 1935. Hydrogen-transferring coenzyme, its composition and mode of action. Biochem. Z. 282:157.

Wick, M. M., A. Rossini, and D. Glynn. 1977. Reduction of streptozotocin toxicity by 3-0-methyl-D-glucose with enhancement of antitumor activity in murine L-1210 leukemia. Cancer Res. 37:3901.

Wilander, E., and R. Gunnarsson. 1975. Diabetogenic effects of n-nitromethylurea in the Chinese hamster. Acta Pathol. Microbiol. Scand. 83:206.

Winter, S. L., and J. L. Boyer. 1973. Hepatic toxicity from large doses of vitamin B-3 nicotinamide. N. Engl. J. Med. 289:1180.

Yoshino, G., T. Kazumi, S. Morita, N. Kobayashi, K. Terashi, and S. Baba. 1979. Glucagon secretion during development of insulin-secreting tumors induced by streptozotocin and nicotinamide. Endocrinol. Jpn. 26:655.

Zackheim, H. S. 1978. Topical 6-amino-nicotinamide plus oral nicotinamide therapy for psoriasis. Arch. Dermatol. 114:1632.

Riboflavin (Vitamin B$_2$)

Riboflavin was first isolated from egg white and was later isolated from milk and liver. The name was adopted after the compound was shown to contain the sugar alcohol, ribitol (Figure 12). It is very slightly soluble in water (11 mg/100 ml) at a neutral or acid pH, but is highly soluble in alkaline solution. Riboflavin is synthesized by the gut microflora and, thus, may reduce the dietary need for the vitamin. Riboflavin is present at various concentrations in a wide range of feeds. It is present in significant quantities in fresh pasture but is low in cereals where it is complexed with proteins. Riboflavin supplements to diets are light-sensitive and may be subject to loss during storage.

NUTRITIONAL ROLE

Dietary Requirements of Various Species

Riboflavin is required in the diets of nonruminant animals. Ruminants obtain sufficient amounts of the vitamin from that synthesized by the rumen microflora. Deficient intakes of the vitamin result in impaired growth. Chronic riboflavin deficiency can be fatal. Specific signs of riboflavin deficiency include seborrheic dermatitis, cheilosis, conjunctivitis, and congenital malformations in rats and mice; curled toe paralysis, reduced embryonic survival, and dermatitis in poultry; muscular weakness, ataxia, dermatitis, anemia, and cardiological changes in dogs and foxes; dermatitis, alopecia, ataxia, corneal degeneration, hemmorhagic adrenals, and fatty degeneration of the kidneys in pigs; and dermatitis and normocytic hypochromic anemia in primates (Nutrition Foundation, 1984). The riboflavin requirements for most species range from 3 to 7 mg/kg of diet.

Biochemical Functions

Riboflavin functions in the intermediary transfer of electrons in metabolic oxidation-reduction reactions as two coenzymes, flavin mononucleotide (FMN) and flavin adenine dinucleotide (FAD). The riboflavin coenzymes function with a large number of oxidases and dehydrogenases important in normal metabolism. Those enzymes that use FMN include glucose oxidase, L-amino acid oxidase, and lactate dehydrogenase. Those that use FAD include D-amino acid oxidase, cytochrome reductase, succinic dehydrogenase, the acyl-CoA dehydrogenases, L-gulonolactone dehydrogenase, α-glycerophosphate dehydrogenase, and glutathione reductase. The activity of the last enzyme in the erythrocyte responds directly to changes in nutritional riboflavin status and is, therefore, used as a clinical parameter for that purpose.

The riboflavin coenzymes transfer electrons to the pyridine dinucleotides of the mitochondrial electron transport chain. Due to this role in energy metabolism, deficient intakes of riboflavin result in impaired efficiency of respiratory energy production. This may result in increases in feed intake by 10 to 15 percent. Reduced electron transport in riboflavin deficiency also results in specific pathologies in those tissues with the greatest normal respiratory rates.

FORMS OF THE VITAMIN

The dietary form of this vitamin may be riboflavin or the coenzyme forms FMN and FAD, which are the predominant forms in mammalian tissues. The supplementary form for addition to diets is generally riboflavin, although some researchers have used sodium riboflavinate. This form is more soluble in water than riboflavin (2 percent versus 11 mg/100 ml).

Riboflavin (vitamin B_2)

FMN

FAD

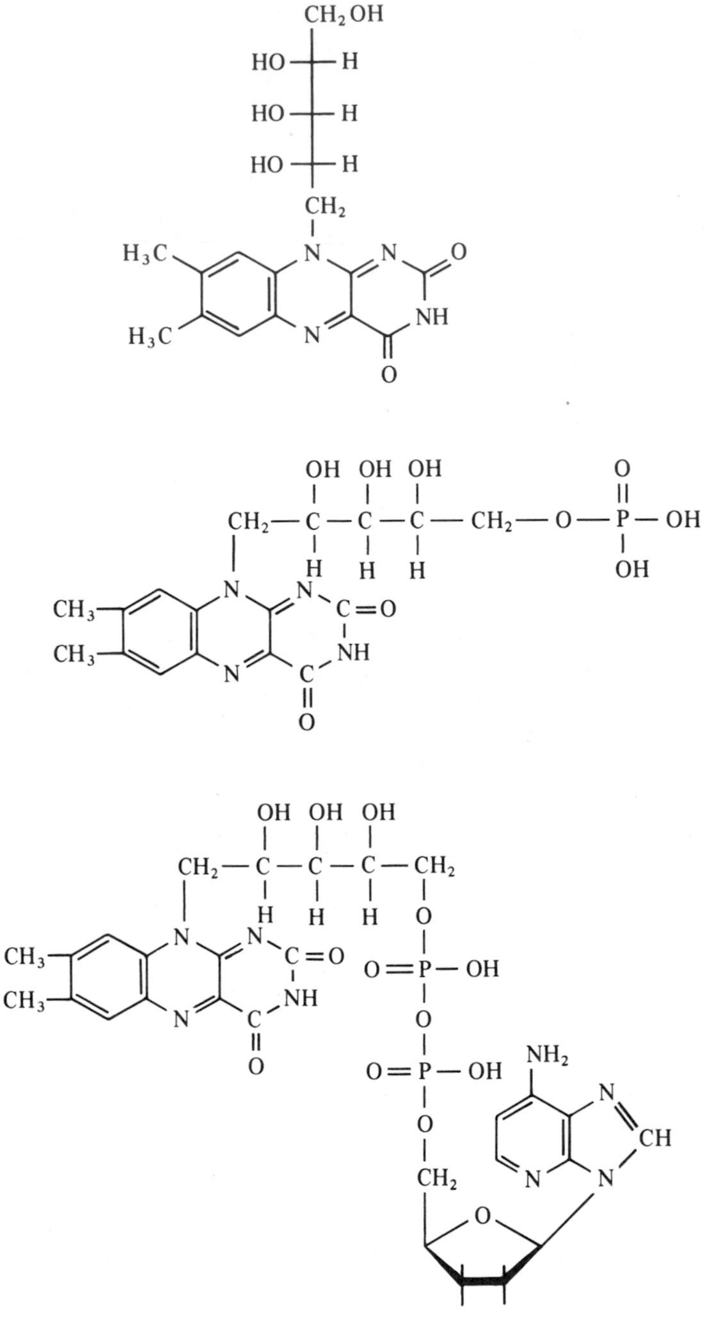

FIGURE 12 Chemical structures of riboflavin and its coenzyme forms.

ABSORPTION AND METABOLISM

Riboflavin is synthesized by the intestinal flora, particularly in ruminant animals. Microbial synthesis of the vitamin takes place in the rumen and/or large intestine, including the cecum. The transit time of the food affects absorption. However, it is not known how much riboflavin can be absorbed in nonruminants. Riboflavin in the circulation is bound to proteins, including immunoglobulins.

Absorption of this water-soluble vitamin occurs in the small intestine. There appears to be little storage in the body. The vitamin is excreted rapidly in the urine, which accounts for the regular requirement for riboflavin by animals. Levels of riboflavin in excess of the requirement in the diet do not appear to be well absorbed, at least in dogs and rats. Unna and Greslin (1942) reported that only 0.1 percent of a dose of riboflavin (exact form not stated) was recovered in the urine of treated dogs within 24 hours following administration. The feces of rats fed large quantities of riboflavin showed an intense yellow color. Massive deposits of riboflavin were reported at the site of subcutaneous injections for as long as 10 days (Unna and Greslin, 1942). Middleton and Grice (1964) studied riboflavin absorption in the intact rat from an orally administered dose of 10 μg and reported recoveries of about 27 percent in the urine and 10 percent in the feces. Axelson and Gibaldi (1972) estimated that about 20 percent of a 1,000-μg oral dose of riboflavin was absorbed in the rat. About 4 to 10 percent was recovered in the urine. When the dose was administered intraperitoneally, 47 to 51 percent was recovered.

HYPERVITAMINOSIS

A summary of the effects of riboflavin administration in animals is shown in Table 13. Seymour et al. (1968) conducted a series of experiments in which 5-week-old pigs were fed diets containing up to 8.8 mg riboflavin/kg at environmental temperatures ranging from -4°C to 32°C for 5 weeks. The minimum riboflavin requirement for maximal BW gain or efficiency of feed conversion was estimated at 3 to 4 mg/kg, with some evidence that it might be increased at low environmental temperatures. Unna and Greslin (1942) found that 10-week-old dogs tolerated oral doses of 24 mg of riboflavin/kg of BW over a 5-month period. No adverse effects on growth or histopathology were reported. Unna and Greslin (1942) reported that the dog could tolerate oral doses of 2 g riboflavin/kg of BW with no evidence of toxicity.

Riboflavin toxicity has been studied in laboratory animals. Ellis et al. (1943) fed diets containing 3 to 10 mg riboflavin/kg to successive generations of rats from 28 to 700 days old. A dietary level of 3 mg/kg appeared to be adequate for breeding stock. No adverse effects on growth or reproduction were noted for higher levels of the vitamin. In fact, the level of 10 mg/kg was found to give the progeny slightly improved growth. Leclerc (1979a) fed diets containing 1 to 16 mg riboflavin/kg to lactating rats and reported an increased tissue concentration with a plateau at 8 mg/kg of diet for dams and 4 mg/kg for pups. Again, no adverse effects were noted at any level. Burch et al. (1956) fed weanling rats diets containing 0 to 15 mg riboflavin/kg for a 5- to 6-week period. They found that a level of 15 mg/kg resulted in increased FMN and FAD levels in the liver, kidney, and heart, but no adverse effects were noted.

Unna and Greslin (1942) administered an oral dose of 10 mg of riboflavin/day for 140 days to weanling rats in each of three successive generations. They reported no adverse effects on growth or reproduction. Doses of up to 5 g/kg of BW were administered by various routes in the same investigation. It was reported that the LD_{50} for riboflavin or sodium riboflavinate administered orally was more than 10 g/kg of BW. For riboflavin or sodium riboflavinate administered by the intraperitoneal route, the LD_{50} was 0.56 g/kg of BW; when administered by the subcutaneous route it was 5 g/kg of BW for riboflavin and 0.79 g/kg of BW for sodium riboflavinate. Unna and Greslin, (1942) found that the rat tolerated oral doses of 10 g of riboflavin or sodium riboflavinate/kg of BW with no evidence of toxicity.

Leclerc (1979b) fed diets containing 1, 2, 4 or 6 mg riboflavin/kg to gestating rats and reported larger litter sizes at levels greater than 1 mg/kg. No effects on progeny whole body riboflavin pool or on dam BW, liver riboflavin concentration, or weight gain were noted. About 30 percent of the riboflavin consumed during gestation was recovered in the urine of each group, although this fell to about 20 percent just before parturition.

Schumacher et al. (1965) fed 0.4 or 10.4 mg of riboflavin/100 g of diet to female rats for 2 weeks prior to mating and through gestation and lactation. They reported that the reproduction of the high-riboflavin group was reduced significantly from 68 percent to 38 percent. These results have not been confirmed by other studies. Mean birth weight, number of pups in a litter, and mean pup BW at weaning were not affected significantly. Concentrations of the vitamin were also not affected in the fetal carcasses or in pup livers at weaning.

PRESUMED UPPER SAFE LEVELS

Insufficient data are available to support estimates of the maximum dietary tolerable levels. Because riboflavin does not appear to be well absorbed from the gut, it is unlikely to present a hazard to animals when included at high levels in the diet.

Leclerc (1979a) reported that a dietary level exceeding 8 mg of riboflavin/kg in the rat (about 3 times the dietary requirement) resulted in a plateauing of the tissue contents of the vitamin. Other available data suggest that levels between 10 and 20 times the dietary requirement can be tolerated safely by rats.

Riboflavin is more toxic when administered parenterally than when administered orally. Estimates of the rat LD_{50} for the intraperitoneal, subcutaneous, and oral routes are 0.56 g/kg, 5 g/kg, and more than 10 g/kg of BW, respectively.

TABLE 13 Research Findings of High Levels of Riboflavin in Animals

Species and No. of Animal	Age or Weight	Administration Amount	Form	Duration	Route	Effect	Reference
Dogs, 4	10 wk	25 mg/kg BW/d	Riboflavin	5 mo	Oral	No adverse effect on growth; no toxic signs; histology normal	Unna and Greslin, 1942
Dogs, 3		2 g/kg BW	Riboflavin	Single dose	Oral	No evidence of toxicity; 0.1% of dose recovered in urine during 24 h	Unna and Greslin, 1942
Swine	4 wk	1.1–8.8 mg/kg	Riboflavin	5 wk	Diet	Growth reduced with less than 2.0 mg/kg	Seymour et al., 1968
Rats, 10/group	12 w	1–6 mg/kg diet	Riboflavin	Gestation	Diet	Dam body and liver concentration increased; no effect in progeny; no effects on dam weight gain or liver weight; litter size improved; urinary excretion not markedly different	Leclerc, 1979b
Rats, 80/group	28 d	3–10 mg/kg diet	Riboflavin	672 d	Diet	Growth slightly improved; no adverse effects on reproduction	Ellis et al., 1943
Rats	Lactating, 0–18 d	1–16 mg/kg diet	Riboflavin	18 d	Diet	Increased tissue concentration with plateau above 8 mg/kg diet in dams, above 4 mg/kg in pups	Leclerc, 1979a
Rats	Weanling, 45 g	0–15 mg/kg diet	Riboflavin	5–6 wk	Diet	Increased FMN and FAD levels in liver, kidney, and heart; no adverse effect on growth	Burch et al., 1956
Rats, 15/group	300 g	10.4 mg/100 g diet	Riboflavin	2 wk prior to mating through lactation	Diet	Litter productivity 38% (68% with controls); no effect on survival or growth of pups; fetal carcass storage unaffected	Schumacher et al., 1965
Rats, 70/group	3 Generations of weanlings	10 mg/d	Riboflavin	140 d	Oral	No adverse effects on growth or reproduction	Unna and Greslin, 1942
Rats		Up to 300 mg/kg BW	Riboflavin	Single dose	IP	LD_{50}, 0.56 g/kg BW; death within 2–5 d	Unna and Greslin, 1942
Rats		Up to 300 mg/kg BW	Sodium riboflavinate	Single dose	IP	LD_{50}, 0.56 g/kg BW	Unna and Greslin, 1942
Rats, 400		Up to 5 g/kg BW	Riboflavin	Single dose	SC	No evidence of toxicity; LD_{50}, 5 g/kg BW	Unna and Greslin, 1942
Rats, 400		Up to 5 g/kg BW	Sodium riboflavinate	Single dose	SC	LD_{50}, 0.79 g/kg BW	Unna and Greslin, 1942
Rats, 400		Up to 10 g/kg BW	Riboflavin or sodium riboflavinate	Single dose	Oral	No evidence of toxicity; LD_{50}, >10 g/kg BW	Unna and Greslin, 1942

SUMMARY

1. Riboflavin is a water-soluble vitamin that is not absorbed well. It is essential in the diets of nonruminant animals.

2. A level exceeding 8 mg of riboflavin/kg of diet (about 3 times the nutritional requirement) results in a plateauing of the tissue contents of the vitamin in rats. Other available data with this species suggest that dietary levels between 10 and 20 times the requirement (possibly 100 times) can be tolerated safely.

3. Riboflavin administered parenterally is more toxic than when given orally. Estimates of the rat LD_{50} for the intraperitoneal, subcutaneous, and oral routes are 0.56 g/kg, 5 g/kg, and more than 10 g/kg of BW, respectively.

REFERENCES

Axelson, J. E., and M. Gibaldi. 1972. Absorption and excretion of riboflavin in the rat—An unusual example of nonlinear pharmacokinetics. J. Pharm. Sci. 61:404.

Burch, H. B., 0.H. Lowry, A. M. Padilla, and A. M. Combs. 1956. Effects of riboflavin deficiency and realimentation on flavin enzymes of tissues. J. Biol. Chem. 223:29.

Ellis, L. N., A. Zmachinsky, and H. C. Sherman. 1943. Experiments upon the significance of liberal levels of intake of riboflavin. J. Nutr. 25:153.

Leclerc, J. 1979a. Influence of the dietary supply of riboflavin on the vitamin nutritional status of the lactating rat and the litter. Int. J. Vit. Nutr. Res. 49:276.

Leclerc, J. 1979b. Vitamin B_2 nutritional status of pregnant rats and their offspring in relation to nutritional intake of riboflavin. J. Int. Vitaminol. Nutr. 49:51.

Middleton, E. J., and H. C. Grice. 1964. Vitamin absorption studies. IV. Site of absorption of C^{14}-riboflavin and S^{35}-thiamine in the rat. Can. J. Biochem. 42:353.

Nutrition Foundation. 1984. P. 285 in Present Knowledge in Nutrition. 5th ed. Washington, D.C.: Nutrition Foundation.

Schumacher, M. F., M. A. Williams, and R. L. Lyman. 1965. Effect of high intakes of thiamine, riboflavin and pyridoxine on reproduction in rats and vitamin requirements of the offspring. J. Nutr. 86:343.

Seymour, E. W., V. C. Speer, and V. W. Hays. 1968. Effect of environmental temperature on the riboflavin requirement of young pigs. J. Anim. Sci. 27:389.

Unna, K., and J. G. Greslin. 1942. Studies on the toxicity and pharmacology of riboflavin. J. Pharmacol. Exp. Ther. 76:75.

Vitamin B$_6$ (Pyridoxine)

The term vitamin B$_6$ is the generic description for the 2-methylpyridine derivatives that have the biological activity of pyridoxine (Figure 13). The vitamin includes aldehyde (pyridoxal) and amine (pyridoxamine) forms. Originally part of Goldberger's "pellagra-preventative factor," vitamin B$_6$ was recognized to have a specific role in preventing dermatitic acrodynia in rats. It was subsequently isolated and identified in the late 1930s. Vitamin B$_6$ is obtained from both plant and animal sources. It is also synthesized by the gut microflora, though in nonruminant animals this source is of doubtful significance.

NUTRITIONAL ROLE

Dietary Requirements of Various Species

Nonruminant animals require a dietary source of pyridoxine to prevent the development of several deficiency signs. These include reduced growth, muscular weakness, hyperirritability, epileptiform convulsions, anemia, acrodynia, scaly dermatitis, and alopecia (NRC, 1978). Nutritional requirements range from 0.9 to 6 mg/kg of diet. Protein intake affects vitamin B$_6$ requirements. Consequently, requirements are often expressed in protein intake terms.

Biochemical Functions

The biologically active forms of vitamin B$_6$ are the coenzymes, pyridoxal phosphate (PLP) and pyridoxalamine phosphate (PMP). PLP is involved in most reactions of amino acid metabolism including transamination, decarboxylation, desulfhydration and nonoxidative deamination. PLP also has roles in the biosynthesis of porphyrins (as a coenzyme for δ-aminolevu-linate synthase) and in the catabolism of glycogen (as part of glycogen phosphorylase). Another role, presently not understood, is apparent in the metabolism of lipids. PLP is important in the metabolism of γ-aminobutyric acid in the brain and in the synthesis of epinephrine and norepinephrine from either phenylalanine or tyrosine.

FORMS OF THE VITAMIN

The predominant dietary form of this vitamin is generally pyridoxine (PN), which is the main form in plant products. Pyridoxal (PL) and pyridoxamine are the principal forms found in animal tissues. All three forms are converted in the animal body to the metabolically active form, PLP. The synthetic form of pyridoxine used for dietary supplementation is generally pyridoxine hydrochloride (PN·HCl) although some researchers have used the free base.

ABSORPTION AND METABOLISM

The rumen microflora synthesize pyridoxine in amounts normally sufficient to meet the needs of ruminants. Microbial synthesis also occurs in the colons of nonruminants. Pyridoxine from this source is not absorbed in appreciable amounts from that organ, however. Absorption of this water-soluble vitamin occurs in the small intestine by a passive process. There appears to be little storage in the body. Differences have been reported in the efficiency of absorption and retention of this vitamin among species. Following an administered dose of PN, the amount recovered in urine was 50 to 70 percent for the rat (Cox et al., 1962), 20 percent for the dog (Scudi et al., 1940), and less than 10 percent for

58

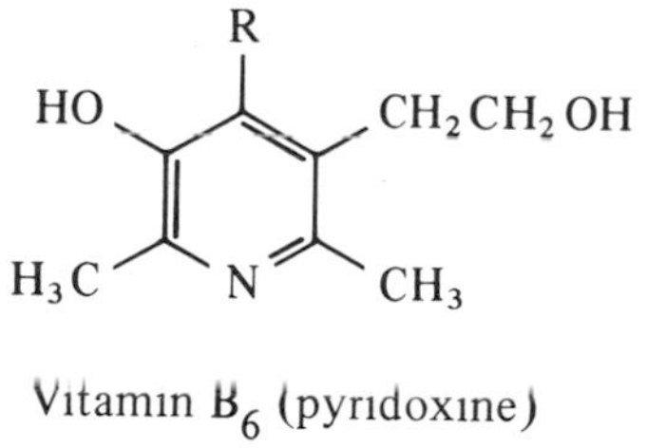

R
|
HO — [structure] — CH$_2$CH$_2$OH

H$_3$C — N — CH$_3$

Vitamin B$_6$ (pyridoxine)

FIGURE 13   General chemical structure of vitamin B$_6$ (pyridoxine). The R group may be CH$_2$OH (pyridoxol), CHO (pyridoxal), or CH$_2$NH$_2$ (pyridoxamine).

humans (Cohen et al., 1973). Pyridoxine is relatively more toxic than other water-soluble vitamins when included in the diet at levels much higher than the nutritional requirement. A main reason for its toxicity is that pyridoxine's passive absorption allows the uptake of massive doses, unlike a saturable absorption mechanism such as that of riboflavin. Consequently, pyridoxine has an acute oral LD$_{50}$ value greater than that of riboflavin.

Once ingested, pyridoxine must be converted to its active forms, PLP and PMP. The conversion requires flavomononucleotides (FMN), flavine adenine dinucleotide (FAD), and niacinamide adenine dinucleotide (NAD). Therefore, a deficiency of niacin or riboflavin necessary for the formation of NAD and FAD, respectively, can result in decreased levels of the active forms of pyridoxine. Pyridoxal phosphate functions with kynureninase in the synthesis of niacin from tryptophan. In pyridoxine deficiency, the diminution of this reaction results in the formation of xanthurenic acid, which is excreted in the urine. Urinary xanthurenic acid is therefore a sensitive indicator of pyridoxine deficiency. About 70 percent of the vitamin is excreted in the urine as the inactive metabolite 4-pyridoxic acid.

HYPERVITAMINOSIS

The toxicity of pyridoxine has been studied in several investigations (Table 14). Adams et al. (1967) fed diets containing 4.8 or 9.2 mg of PN (PN·HCl) to 7-kg early weaned pigs for 122 days and reported better growth and feed efficiency with the higher level of supplementation. Dogs given oral doses of 20 mg/kg of BW for 75 days did not develop any toxic signs (Unna and Antopol, 1940). Phillips et al. (1978) administered higher oral doses of 50 mg of PN·HCl/kg of BW/day and reported no signs of toxicity.

Higher doses of the vitamin have been found to produce signs of toxicity. Phillips et al. (1978) reported that ataxia, muscle weakness, and loss of balance developed between 40 and 75 days in dogs that received 200 mg of

PN·HCl/kg of BW/day. Dogs fed daily doses of 250 mg/kg/day began to develop incoordination and ataxia within the first week of treatment. The dose was then reduced to 200 mg/kg of BW/day for the remainder of the experimental period. Histological examination of the tissues revealed bilateral loss of myelin and axons in the dorsal funiculi and loss of myelin in individual fibers of the dorsal nerve roots. A lesser amount of pathological damage was observed in dogs receiving 50 mg/kg of BW/day. Analyses of tissues revealed elevated concentrations of PN in the blood, cerebral cortex, spinal cord, spleen, kidney, and muscle of animals receiving 200 mg/kg/day. The group receiving 50 mg/kg of BW/day had elevated PN levels only in the blood and cerebral cortex. Schaeppi and Krinke (1982) and Antopol and Tarlov (1942) administered oral doses of 1.5 or 3 g of PN·HCl/kg of BW/day to dogs weighing about 10 kg for periods of up to 26 days. Toxicity was noted after 2 days. Histological lesions of the sensory neurons and spinal column were recorded.

Hoover and Carlton (1981) administered daily doses of PN·HCl to beagle dogs according to a regimen that raised the dose from 50 to 150 mg/kg of BW by the fifteenth day and continued at that level for 85 days. The pyridoxine treatment produced anorexia within 2 weeks and ataxia within 4 weeks. Krinke et al. (1980) administered daily oral doses of 300 mg of PN·HCl/kg of BW to pairs of 7- to 11-month-old beagle dogs for 78 days. They reported the development of a locomotory abnormality (swaying gait) within 9 days. Treated dogs eventually became unable to walk, but did not show muscular weakness. It was concluded that pyridoxine produced a toxic, peripheral, sensory neuronopathy involving degeneration of the dorsal root ganglia, gasserian ganglia, and sensory nerve fibers.

Workers have fed diets containing up to 1,430 mg of PN·HCl per kg to growing and breeding rats over prolonged periods with no adverse effects (Brin and Thiele, 1967; Cohen et al., 1973; Stowe et al., 1974; Alton-Mackey and Walker, 1978; Kirksey and Susten, 1978; Sloger and Reynolds, 1980; Mercer et al., 1984). In addition, daily oral doses of up to 2.5 mg of PN·HCl/rat over a prolonged period, or oral doses of 9 mg of PN·HCl given on each of 2 successive days did not result in any adverse effects (Unna and Antopol, 1940).

When Erabi et al. (1983) injected PLP into the ventricular sinus of rats, the animals exhibited convulsions. Weigand et al. (1940) administered single intravenous doses of 300 to 700 mg of PN·HCl/kg of BW to mice or rats. They reported mortality in mice with doses higher than 300 mg/kg and in rats with doses higher than 500 mg/kg. The acute LD$_{50}$ value for mice was estimated to be 545.3 mg/kg. For rats it was 657.5 mg/kg. Schumacher et al. (1965) fed diets containing 2.5 (control) or

TABLE 14 Research Findings of High Levels of Vitamin B_6 (Pyridoxine) in Animals

Species and No. of Animal	Age or Weight	Administration Amount	Form	Duration	Route	Effect	Reference
Dogs, 3	Pups	20 mg/kg BW		75 d	Oral	Growth unaffected; no adverse effects on blood parameters; histology normal	Unna and Antopol, 1940
Dogs, beagles, 5/treatment	7–8 mo (8 kg)	50 mg/kg BW/d	PN · HCl	107 d	Oral	No clinical effects; reduction of myelin in nerves; increased PN concentration in blood and cerebral cortex	Phillips et al., 1978
Dogs, male and female beagles, 10–13/group	13–15 mo	50 increasing to 150 mg/kg BW/d	PN · HCl	100 d	Oral	Anorexia within 2 wk; ataxia within 4 wk	Hoover and Carlton, 1981
Dogs, beagles, 5/treatment	7–8 mo (8 kg)	200 mg/kg BW/d	PN · HCl	107 d	Oral	Ataxia, muscle weakness, and loss of balance after 40–75 d; reduction of myelin in nerves; increased PN concentration in blood and organs	Phillips et al., 1978
Dogs, beagles, 5/treatment	7–8 mo (8 kg)	250 mg/kg BW/d	PN · HCl	1 wk	Oral	Incoordination; ataxia	Phillips et al., 1978
Dogs, male and female beagles, 2/group	7–11 mo	300 mg/kg BW/d	PN · HCl	78 d	Oral	Development of swaying gait within 9 d; neuronal degeneration of ganglia	Krinke et al., 1980
Dogs, beagles, 2/treatment	12 mo (11 kg)	3 g/d	PN · HCl	8–26 d	Oral	Unsteady gait; weakness; apathy; neurologic impairment. lesions of sensory neurons	Schaeppi and Krinke, 1982
Dogs, males and females, 4 total	Adult (~10 kg)	1–5 g/kg BW/d	PN or PN · HCl	1–4 d	Oral	Vomiting after dosing; ataxia after 2 d. Sacrificed or died at 8–14 d; degeneration of posterior columns of spine found	Antopol and Tarlov, 1942
Humans, 7	20–43 yr	2–6 g/d	PN · HCl	3–40 mo	Oral	Ataxia; sensory and nervous system dysfunction; 4 humans severely disabled	Schaumburg et al., 1983
Mice, 3–5/group		300–700 mg/kg BW	PN · HCl	Single dose	IV	LD_{50}, 545.3 mg/kg; no mortality with 300 mg/kg	Weigand et al., 1940
Rats, 10		1 g/kg BW	PN (free base)	1 d	Oral	Tonic convulsions after 24 h; LD_{50}, 4 g/kg	Unna and Antopol, 1940
Rats, males, 13	Adult	3–7 g/kg BW	PN (free base)	Single dose	Oral	Movements uncoordinated after 24 h; sacrificed at 4–37 d; rarefaction of posterior columns of spinal cord found	Antopol and Tarlov, 1942
Rats, 10/group	120–150 g	1–8 g	PN or PN · HCl	Single dose	SC	LD_{50}, 3.1 g/kg for PN; 3.7 g/kg for PN · HCl	Unna, 1940
Rats, males, 2	Adult	5 g/kg BW	PN · HCl	Single dose	Oral	Movements uncoordinated after 24 h; sacrificed at 4 d; no histological effects found	Antopol and Tarlov, 1942
Rats, 15	3 wk	2.5 mg/d	PN · HCl	87 d	Oral	No adverse effects on growth; no abnormalities at autopsy	Unna and Antopol, 1940
Rats, 6	87 d–maturity	2.5 mg/d	PN · HCl		Oral	Two litters obtained (5–8 pups); mean birth weight 5 g	Unna and Antopol, 1940
Rats, males, 42	112 g	0, 0.25, 0.5, 1.0, 2.0, 4.0, and 10 mg/kg diet (purified diet lacking PN fed for prior 2 weeks)	PN · HCl	21 d	Diet	Daily feed intake and weight gain leveled off at about 1 mg/kg diet; no depression at 10 mg/kg; serum glutamic pyruvic transaminase activity increased to 10 mg/kg	Mercer et al., 1984

Species, number	Body weight/age	Dose	Form	Duration	Route	Effects	Reference
Rats, 8 (PN deficient)	40–50 g	9 mg/d	PN · HCl	2 d	Oral	Prompt cure; no adverse effects	Una and Antopol, 1940
Rats, males, 6–7/group	Weanling	Up to 10 mg/kg diet	PN · HCl	14 d	Diet	No adverse effects on growth; increased content in tissues	Bein and Thiele, 1967
Rats, female, Sprague-Dawley, 10/group	120 d	Up to 19.2 mg/kg diet	PN · HCl	Throughout gestation—d 12 of lactation	Diet	No adverse effects on reproduction; increased content in tissues and milk; saturation at 2.4 mg/kg (muscle), 4.8 mg/kg (liver), and 9.6 mg/kg (milk)	Kirksey and Susten, 1978
Rats, male and female, Sprague-Dawley, 6–12/group	Birth	Up to 19.2 mg/kg diet	PN · HCl	13 wk	Diet	No adverse effects on growth	Kirksey and Susten, 1978
Rats, female, Wistars, 4/group		24 mg/kg diet	PN · HCl[a]	21 d	Diet	No adverse effect on BW of pups; improved neuromotor skills	Alton-Mackey and Walker, 1978
Rats	300 g	0.25, 6.25 mg/100 g diet	PN[a]	2 wk prior to mating through lactation	Diet	Litter productivity 47% (68% in controls); no effect on survival or growth of pups; fetal carcass storage of PN increased from 0.79 to 1.23 μg/g	Schumacher et al., 1965
Rats, male, Nelson-Wistars, 26 pairs	50–60 g	500 mg/kg diet	PN · HCl	5 wk	Diet	Increased weight gain, liver weight; improved feed efficiency; no obvious physical differences	Cohen et al., 1973
Rats, male and female, Charles River CD strain, 4–7/group	56 d	1,430 mg/kg diet	PN · HCl	45 d	Diet	No adverse effect on feed intake, growth rate, or reproduction; PLP in plasma of unmated females increased; no increase in dams or pups. Blood PLP increased in dams and pups	Slager and Reynolds, 1980
Rats, female, Wistars	175–200 g, mated	20–80 mg/kg/d	PN · HCl	Single dose on 6–15 of gestation	Oral	No adverse effects on fetal mortality, fetal weight, resorption sites or fetal anomalies at d 22 of gestation	Khera, 1975
Rats, 15/group	30 g	0.5–2.5 mg/d	PN[a]	80 d	Oral	No effect on growth	Una, 1940
Rats, male	150 g	300 mg/kg BW	PN · HCl	31 d	Oral	Slight ataxia	Krinke et al., 1978
Rats, female, Sprague-Dawley, 6	Mated	100 mg/100 g BW	PN · HCl	Single treatment on d 21; killed 5 h later	IP	Apparent stimulation of TTA and HTA in dams and fetuses; no adverse effects on reproduction	Susten and Kirksey, 1970
Rats, 5–10/group		500–700 mg/kg BW	PN · HCl	Single dose	IP	LD$_{50}$, 657.5 mg/kg; no mortality with 500 mg/kg	Weigand et al., 1940
Rats, 10		1 g/kg BW	PN · HCl	1 d	SC	LD$_{50}$, 3.7 g/kg; tonic convulsions after 24 h	Una and Antopol, 1940
Rats, 10		1 g/kg BW	PN (free base)	1 d	SC	LD$_{50}$, 3.1 g/kg; tonic convulsions after 24 h	Una and Antopol, 1940
Rats, 10		1 g/kg BW	PN · HCl	1 d	Oral	LD$_{50}$, 6 g/kg; tonic convulsions after 24 h	Una and Antopol, 1940
Swine, 48/group	Weanling (7 kg)	9.2 mg/kg diet	PN[a]	122 d	Diet	Growth and feed efficiency better than with 4.8 mg/kg	Adams et al., 1967

[a] It is presumed that this form was used.

62.5 mg of PN/kg ad libitum to female rats from 2 weeks before mating through gestation and lactation. They reported that the reproduction of the high-PN group was reduced by 47 percent versus 68 percent in the control group. Other investigations have not confirmed this result. Mean birth weight, mean number of pups per litter, and mean pup weight at weaning were not affected significantly. The higher level of dietary PN resulted in a significant increase of 0.79 to 1.23 μg in the carcass PN content of the newborns. The level did not affect the PN requirement of the weanling animals. Krinke et al. (1978) administered an oral dose of 300 mg of PN·HCl/kg of BW to male rats for 31 days. They observed a slight ataxia. A single intraperitoneal dose of 1,000 mg/kg of BW administered to dams the twenty-first day of gestation had no adverse effects on reproduction (Susten and Kirksey, 1970). There were apparent stimulations of tyrosine transaminase and holotyrosine transaminase activities in both dams and fetuses. Single oral or subcutaneous doses of up to 7 g/kg of BW were administered in the investigations of Antopol and Tarlov (1942) and Unna and Antopol (1940). After 24 hours, workers noted uncoordinated movement and tonic convulsions. They observed rarefaction of the posterior columns of the spinal cord. The LD_{50} values for PN·HCl in rats were determined to be 6 g/kg of BW by the oral route and 3.7 g/kg of BW by the subcutaneous route. The corresponding values for PN were 4 and 3.1 g/kg of BW. The 19 percent difference in the acute toxicity by the subcutaneous route between these two forms of the vitamin is consistent with the 18 percent difference in their molecular weights. This suggests that the toxicity is attributable to the pyridoxine portion rather than to the hydrochloride component.

Unna (1940) administered subcutaneous and oral doses of 1 to 8 g of PN or PN·HCl to 120- to 150-g rats and reported that up to 1 g/kg of BW was tolerated without adverse effects. Larger doses resulted in muscular incoordination in 2 or 3 days leading to convulsions and death. By subcutaneous administration, the LD_{50} values were 3.1 g/kg of BW for PN and 3.7 g/kg of BW for PN·HCl. By oral administration, the LD_{50} of PN·HCl was 5.5 g/kg of BW, which suggested that the vitamin in that form was readily absorbed from the gastrointestinal tract. In addition, young 30-g rats given daily oral doses of 0.5 to 2.5 mg of PN for 80 days showed no adverse growth effects.

Pyridoxine has been used at high doses in humans as a treatment for conditions ranging from premenstrual syndrome to schizophrenia. Oral doses of 2 to 6 g/day to adults over a prolonged period are associated with sensory-nervous system dysfunction and disablement (Schaumburg et al., 1983). Pyridoxine has also been used to depress abnormally high lactation in women

(Rose, 1978), possibly by increasing the formation of dopamine. Prolactin was decreased by doses of 200 mg of PN given 3 times a day (Rose, 1978).

PRESUMED UPPER SAFE LEVELS

Insufficient data are available to support estimates of the maximum dietary tolerable levels of vitamin B_6 for species other than the dog and the laboratory rat. Levels of PN of 1,000 mg/kg of diet fed for less than 60 days, or less than 500 mg/kg of diet fed for more than 60 days, appear to be safe for dogs.

The available data suggest that rats may safely be fed diets containing up to 500 mg of PN/kg for less than 60 days, or up to 250 mg of PN/kg for more than 60 days. Estimates of the dietary levels of PN·HCl that produce specific tissue and body fluid saturation in rats following exposure for more than 60 days are: muscle, 2.4 mg/kg; liver, 4.8 mg/kg; milk, 9.6 mg/kg. Estimates of the acute oral LD_{50} for the rat are 3.1 to 4 g/kg for PN and 3.7 to 6 g/kg for PN·HCl. It is suggested that dietary levels of at least 50 times nutritional requirements are safe for most species.

SUMMARY

1. Vitamin B_6 (pyridoxine) is a water-soluble vitamin that is absorbed readily. Some domestic and laboratory animals require a dietary source of the vitamin.

2. Pyridoxine can be toxic to animals when administered at high levels. A level of 1,000 mg of PN·HCl/kg of diet appears safe for dogs. Rats may safely be fed diets containing up to 500 mg of PN/kg for less than 60 days, or up to 250 mg of PN/kg for more than 60 days. Estimates of the dietary levels of PN·HCl that produce specific tissue and body fluid saturation in rats following exposure for more than 60 days are: muscle, 2.4 mg/kg; liver, 4.8 mg/kg; milk, 9.6 mg/kg. Estimates of the acute oral LD_{50} for the rat are 3.1 to 4 g/kg for PN and 3.7 to 6 g/kg for PN·HCl.

3. Available evidence from dog and rat studies suggests that probably more than 1,000 times the nutritional requirements would have to be included in diets in order to produce signs of toxicity in these particular species.

REFERENCES

Adams, C. R., C. E. Richardson, and T. J. Cunha. 1967. Supplemental biotin and vitamin B_6 for swine. J. Anim. Sci. 26:903. (Abstr.)
Alton-Mackey, M. G., and B. L. Walker. 1978. The physical and neuromotor development of progeny of female rats fed graded levels of pyridoxine during lactation. Am. J. Clin. Nutr. 31:76.

Antopol, W., and I. M. Tarlov. 1942. Experimental study of the effects produced by large doses of vitamin B$_6$. J. Neuropathol. Exp. Neurol. 1:330.

Brin, M., and V. F. Thiele. 1967. Relationships between vitamin B$_6$ vitamer content and the activities of two transaminase enzymes in rat tissues at varying intake levels of vitamin B$_6$. J. Nutr. 93:213.

Cohen, P. A., K. Shneidman, F. Ginsberg-Fellner, J. A. Sturman, J. Knittle, and G. E. Gaull. 1973. High pyridoxine diet in the rat: Possible implications for megavitamin therapy. J. Nutr. 103:143.

Cox, S. H., A. Murray, and I. V. Boone. 1962. Metabolism of tritium-labelled pyridoxine in rats. Proc. Soc. Exp. Biol. Med. 109:242.

Erabi, M., D. E. Metzler, and W. R. Christensen. 1983. Convulsant activity of pyridoxal sulphate and phosphoethyl pyridoxal: Antagonism by GABA and its synthetic analogues. Neuropharmacology 22:865.

Hoover, D. M., and W. W. Carlton. 1981. The subacute neurotoxicity of excess pyridoxine HCl and clioquinol (5-chloro-7-iodo-8-hydroxyquinoline) in beagle dogs. 1. Clinical disease. Vet. Pathol. 18:745.

Khera, K. S. 1975. Teratogenicity study in rats given high doses of pyridoxine (vitamin B$_6$) during organogenesis. Experientia 31:469.

Kirksey, A., and S. S. Susten. 1978. Influence of different levels of dietary pyridoxine on milk composition in the rat. J. Nutr. 108:113.

Krinke, G., J. Heid, H. Bittiger, and R. Hess. 1978. Sensory denervation of the plantar lumbrical muscle spindles in pyridoxine neuropathy. Acta Neuropathol. 43:213.

Krinke, G., H. H. Schaumberg, P. S. Spencer, J. Suter, P. Thomann, and R. Hess. 1980. Pyridoxine megavitaminosis produces degeneration of peripheral sensory neurons (sensory neuronopathy) in the dog. Neurotoxicology 2:13.

Mercer, L. P., J. M. Gustafson, P. T. Higbee, C. E. Geno, M. R. Schweisthal, and T. B. Cole. 1984. Control of physiological response in the rat by dietary nutrient concentration. J. Nutr. 114:144.

National Research Council. 1978. Nutrient Requirements of Laboratory Animals, 3rd rev. ed. Washington, D.C.: National Academy Press.

Phillips, W. E. J., J. H. L. Mills, S. M. Charbonneau, L. Tryphonas, G. V. Hatina, Z. Zawidzka, F. R. Bryce, and I. C. Munro. 1978. Subacute toxicity of pyridoxine hydrochloride in the beagle dog. Toxicol. Appl. Pharmacol. 44:323.

Rose, D. 1978. Interactions between vitamin B$_6$ and hormones. Vit. Horm. 36:53.

Schaeppi, U., and G. Krinke. 1982. Pyridoxine neuropathy: Correlation of functional tests and neuropathology in beagle dogs treated with large doses of B$_6$. Agents Actions 12:575.

Schaumburg, H., J. Kaplan, A. Windebank, N. Vick, S. Rasmus, D. Pleasure, and M. J. Brown. 1983. Sensory neuropathy from pyridoxine abuse: A new megavitamin syndrome. N. Engl. J. Med. 309:445.

Schumacher, M. F., M. A. Williams, and R. L. Lyman. 1965. Effect of high intakes of thiamine, riboflavin and pyridoxine on reproduction in rats and vitamin requirements of the offspring. J. Nutr. 86:343.

Scudi, J. V., K. Unna, and W. Antopol. 1940. A study of the urinary excretion of vitamin B$_6$ by a colorimetric method. J. Biol. Chem. 135:371.

Sloger, M. S., and R. D. Reynolds. 1980. Effects of pregnancy and lactation on pyridoxal 5'-phosphate in plasma, blood and liver of rats fed three levels of vitamin B$_6$. J. Nutr. 110:1517.

Stowe, H. D., R. A. Croyer, P. Medley, and M. Cates. 1974. Influence of dietary pyridoxine in cadmium toxicity in rats. Arch. Environ. Health 28:209.

Susten, S. S., and A. Kirksey. 1970. Influence of pyridoxine on tyrosine transaminase activity in maternal and fetal rat liver. J. Nutr. 100:369.

Unna, K. 1940. Studies on the toxicity and pharmacology of vitamin B$_6$ (2-methyl-3-hydroxy-4,5-bis(hydroxymethyl)-pyridoxine). J. Pharmacol. 70:400.

Unna, K., and W. Antopol. 1940. Toxicity of vitamin B$_6$. Proc. Soc. Exp. Biol. Med. 43:116.

Weigand, C. G., C. R. Eckler, and K. K. Chan. 1940. Action and toxicity of vitamin B$_6$ hydrochloride. Proc. Soc. Exp. Biol. Med. 44:147.

Folic Acid

Stokstad (1943) isolated folic acid as a result of investigations of the properties of factors present in yeast or liver that would promote the growth of lactic acid bacteria (Snell and Peterson, 1940). Subsequent studies demonstrated that the active factor was identical to, or related to, antianemia factors and animal growth factors discovered by other investigators (Brody et al., 1984). Mowat et al. (1948) showed that folic acid is composed of a pteridine ring linked through a methylene bridge to *p*-aminobenzoic acid to form pteroic acid, which is in turn linked as an amide to glutamic acid. Early studies of the ability of purines or thymine to partially satisfy bacterial requirements for folic acid pointed to the involvement of this vitamin in purine and thymine biosynthesis. Early studies using ^{14}C-labeled formate and formaldehyde suggested a role of folic acid in the metabolism of one-carbon units.

NUTRITIONAL ROLE

Dietary Requirements of Various Species

The nutritional status of animals with respect to folate adequacy is most often assessed through an estimation of folate concentrations in serum or red cells. These determinations have historically been carried out through microbiological assays, but protein-binding assays are used increasingly. Functional tests are used as well. For example, excretion of formiminoglutamate (FIGLU) following a histidine load is often used as a clinical measure of folate adequacy. Growth rate and the maintenance of normal hematological responses have also been used as a measure of folate adequacy in animal studies.

Folate requirements have been determined to range from 0.25 to 1.0 mg/kg of diet for chickens and 1 to 6 mg/kg of diet for rats and guinea pigs. Sulfa drugs, which are often added to commercial chick diets, increase the folate requirement. Folate requirements for ruminant animals and horses have not been established. The swine requirement is less than 1 mg/kg of diet. The folate requirement for the human is about 50 μg/day and is increased during pregnancy and lactation.

Biochemical Functions

The pteridine ring is completely oxidized in folic acid and can be enzymatically reduced to the dihydro- and tetrahydrofolate forms. The coenzyme forms of the vitamin are the various one-carbon derivatives of the tetrahydrofolate (see Figure 14). These include the 5-methyl-, 5- and 10-formyl-, 5-formimino-, 5,10 methylidene-, and 5-10 methylene- derivatives (Wagner, 1984). They encompass one-carbon units at the oxidation state of methanol, formaldehyde, or formate. These one-carbon derivatives are generated from free formate or from the metabolism of glycine, serine, and histidine. The derivatives can be oxidized and reduced as folate coenzymes. The various one-carbon units carried on folate coenzymes are used to synthesize methionine and purine rings and convert deoxyuridinemonophosphate to deoxythymadinemonophosphate. The folate coenzymes, *S*-adenosyl methionine and vitamin B_{12}, are therefore responsible for the movement of one-carbon units in metabolic pathways.

FORMS OF THE VITAMIN

Folacin is the generic descriptor of all compounds that exhibit the biological activity of folic acid. Pteroylglu-

64

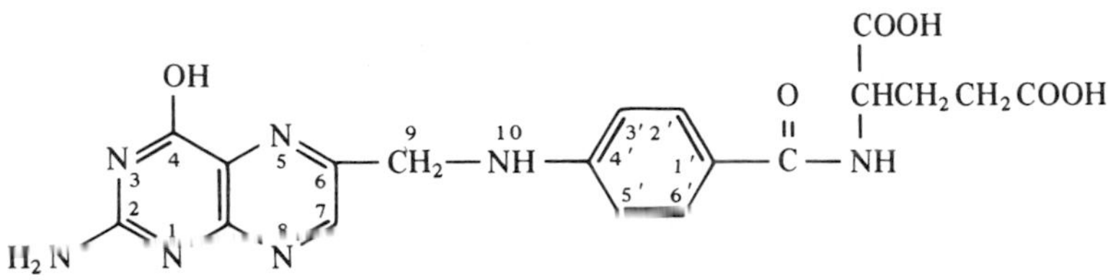

Folic acid (PteGlu)

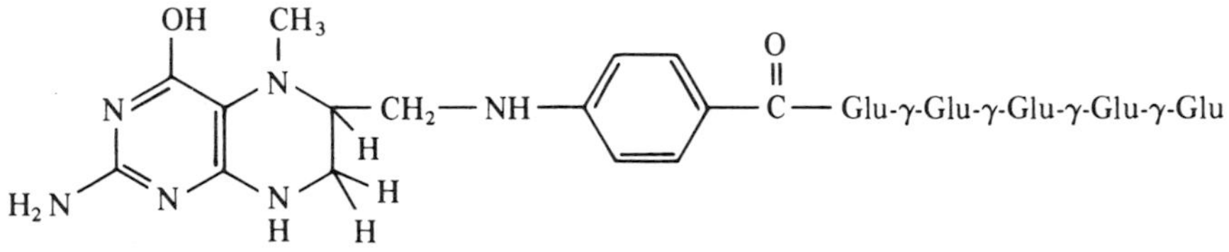

Methyl-tetrahydrofolate pentaglutamate
(5-methyl-H$_4$PteGlu$_5$)

FIGURE 14   Chemical structures of folic acid and 5-methyl-tetrahydrofolate.

tamic acid (2-amino-4-hydroxy-6-methyleneaminio-benzoyl-L-glutamic acid pteridine) was the first form of folate isolated and synthesized. In almost all tissues, however, the predominant forms are polyglutamates. These forms may have up to eight additional glutamic acid residues attached to the terminal glutamate of folic acid in an amide linkage as a poly-γ-amide. The polyglutamates are the reduced active coenzyme forms in animal tissues. The monoglutamate is used in the supplementation of animal feeds. Animals convert the absorbed monoglutamate form to polyglutamates. Subsequent metabolism of the reduced polyglutamates produces the various one-carbon derivatives that function in metabolic reactions.

ABSORPTION AND METABOLISM

Foods contain a mixture of the mono- and polyglutamate forms of folate. These forms are predominantly in the reduced state. The polyglutamates are hydrolyzed by a γ-glutamylhydrolase prior to absorption as the monoglutamate. The occurrence of folate deficiency in celiac sprue is the result of disturbance of the proximal small intestine mucosa and a failure of the normal intestinal transport process. Absorbed monoglutamates are transported in plasma to cells that use specific transport systems to take up the vitamin (Herbert et al., 1980). The majority of body folate occurs in the liver in the form of the 5-methyl- or 10-formyl-tetrahydro derivatives of the penta or hexaglutamate. Both free folate and folate degradation products are excreted in the bile. There is a substantial enterohepatic circulation of the vitamin, which leads to a significant fecal loss.

HYPERVITAMINOSIS

Adverse effects following the ingestion of elevated amounts of folic acid to animals have not been observed. The vitamin is generally regarded as nontoxic. Pharmacological responses to massive doses of folic acid have, however, been reported (Omaye, 1984; Preuss, 1978). Single intravenous doses of 250 mg/kg of BW of sodium folate caused an epileptic response in rats (Hommes and Obbens, 1973). This response is decreased in partially hepatectomized rats. These data suggest that metabolism is required for the toxic response, but the nature of the metabolic change has not been identified. Threlfall et al. (1966) observed that parenteral administration of 250 mg folate/kg of BW causes a renal hypertrophy in the rat. This effect is not seen in the villi of the small bowel (Tilson, 1970), and it is likely that the renal effect is due to tubular obstruction rather than to any metabolic effect. Schmidt and Dubach (1976) and Schubert (1976) have investigated the morphological and metabolic changes associated with the pteridine-induced stimulus of renal growth in some detail.

PRESUMED UPPER SAFE LEVELS

No adverse responses to the ingestion of folic acid have been documented. Therefore, the upper limits of presumed safe dietary levels cannot be established.

SUMMARY

1. Administration of massive doses of folic acid in the diet has not been reported to cause adverse effects.

2. Single parenteral doses of folic acid have been reported to induce epileptic responses and renal hypertrophy in rats. The doses employed to produce this response have been about 1,000 times greater than the dietary requirements for the vitamin.

REFERENCES

Brody, T., B. Shane, and R. Stokstad. 1984. Folic acid. Pp. 459–496 in Handbook of Vitamins, L. J. Machlin, ed. New York: Marcel Dekker.

Herbert, V., N. Colman, and E. Jacob. 1980. Folic acid and vitamin B_{12}. Pp. 229–258 in Modern Nutrition in Health and Disease, R. S. Goodhart and M. E. Shils, eds. Philadelphia: Lea & Febiger.

Hommes, O. R., and E. A. M. T. Obbens. 1973. Liver function and folate epilepsy in the rat. J. Neurol. Sci. 20:269.

Mowat, J. H., J. H. Boothe, B. L. Hutchings, E. L. R. Stokstad, C. W. Waller, R. B. Angier, J. Semb, D. B. Cosulich, and Y. Subbarow. 1948. The structure of liver L. casei factor. J. Am. Chem. Soc. 70:14.

Omaye, S. T. 1984. Safety of megavitamin therapy. Adv. Exp. Med. Biol. 177:169.

Preuss, H. 1978. Effect of nutrient toxicities—Excess in animals and man: Folic acid. Pp. 61–62 in Handbook in Nutrition and Food, M. Rechcigl, ed. New York: CRC Press.

Schmidt, U., and U. C. Dubach. 1976. Acute renal failure in the folate-treated rat: Early metabolic changes in various structures of the nephron. Kidney Int. 10:S39.

Schubert, G. E. 1976. Folic acid-induced acute renal failure in the rat: Morphological studies. Kidney Int. 10:S46.

Snell, E. E., and W. H. Peterson. 1940. Growth factors for bacteria. X. Additional factors required by certain lactic acid bacteria. J. Bacteriol. 39:273.

Stokstad, E. L. R. 1943. Some properties of a growth factor for Lactobacillus casei. J. Biol. Chem. 149:573.

Threlfall, G., D. M. Taylor, and A. T. Buck. 1966. The effect of folic acid on growth and deoxyribonucleic acid synthesis in the rat kidney. Lab. Invest. 15:1477.

Tilson, M. D. 1970. A dissimilar effect of folic acid upon the growth of the rat kidney and small bowel. Proc. Soc. Exp. Biol. Med. 134:95.

Wagner, C. 1984. Folic acid. Pp. 332–346 in Present Knowledge in Nutrition, R. E. Olsen, ed. Washington, D.C.: The Nutrition Foundation.

Pantothenic Acid

Williams (1939) first isolated pantothenic acid as a growth factor for yeast. The vitamin was also investigated as a growth-promoting factor for lactic acid bacteria. Pantothenic acid is widely distributed in biological materials. Subsequent to its isolation as a microbial growth factor, it was shown that pantothenic acid is identical to an antidermatitis factor for chicks and a growth-promoting factor for rats (Fox, 1984; Olson, 1984). The presence of pantothenic acid as a component of coenzyme A (Lipmann et al., 1947) indicated the vitamin's biochemical role and led to the demonstration that pantothenic acid-deficient rats had defects in the ability to metabolize fatty acids.

NUTRITIONAL ROLE

Dietary Requirements of Various Species

Signs of pantothenic acid deficiency vary in different species (Sauberlich, 1980; Fox, 1984). An effect on growth response usually can be demonstrated. Most pantothenic acid-deficient laboratory animals exhibit dermatitis, achromotrichia, and nasal porphyrin excretion. Pantothenic acid-deficient poultry exhibit a feathering disorder, fatty livers, and a characteristic dermatitis in the corners of the mouth.

Deficiency signs are readily produced in most laboratory animals. The dietary requirements for rats, mice, and guinea pigs are 8 to 20 mg/kg of diet; for chicks, 10 mg/kg; for swine, 13 mg/kg; and for dogs and cats, 10 mg/kg. No daily dietary requirements for pantothenic acid have been established for ruminants, horses, or humans. Estimates of human adult daily intakes of the vitamin in the United States range from 5 to 20 mg/day (Sauberlich, 1980).

Biochemical Functions

Pantothenic acid is a component of coenzyme A, acyl CoA synthetase, and acyl carrier protein. The coenzyme form of the vitamin is therefore responsible for acyl group transfer reactions. The acyl derivatives of coenzyme A are activated thiol esters of the β-mercaptoethylamine portion of the molecule, which is attached to the carboxyl group of pantothenic acid to form the active coenzyme. These activated acyl groups are involved in condensations, acyl group exchanges, and acyl group transfers catalyzed by a number of enzymes. Coenzyme A derivatives are also involved in fatty acid degradation, and fatty acids are synthesized as acyl carrier protein derivatives.

FORMS OF THE VITAMIN

Pantothenic acid consists of pantoic acid (α,γ-dihydroxy-β,β'-dimethylbutyric acid) joined to β-alanine by an amide bond (Figure 15). Much of the pantothenic acid in tissues consists of the coenzyme forms of the vitamin. These forms all have β-mercaptoethylamine bound as an amide to pantothenic acid and have a 4'-phosphate joined to a 3',5'-adenosine diphosphate (ADP) by a pyrophosphate in coenzyme A, or to a serine residue of acyl carrier protein or acyl CoA synthetase. This mixture of coenzymes is to a large extent acylated in tissues. Analyses of the vitamin in foods and tissues have most often been carried out by microbiological methods, and enzymatic digestion has been used to liberate pantothenic acid from the various coenzyme forms. The form used in the supplementation of animal feeds is the salt, calcium pantothenate.

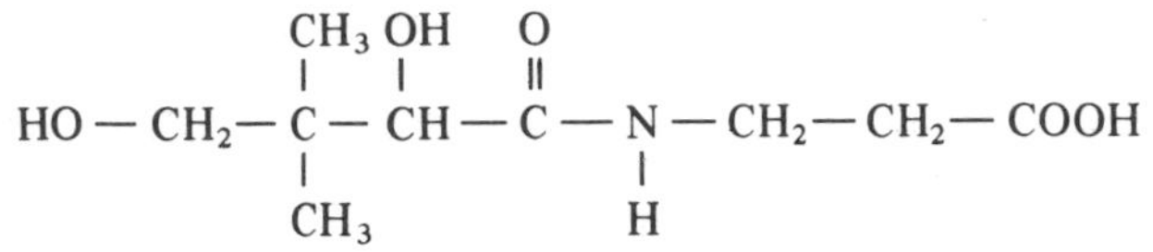

CH₃ OH O structure...

HO — CH₂ — C — CH — C — N — CH₂ — CH₂ — COOH
with CH₃ below C and H below N

Pantothenic acid

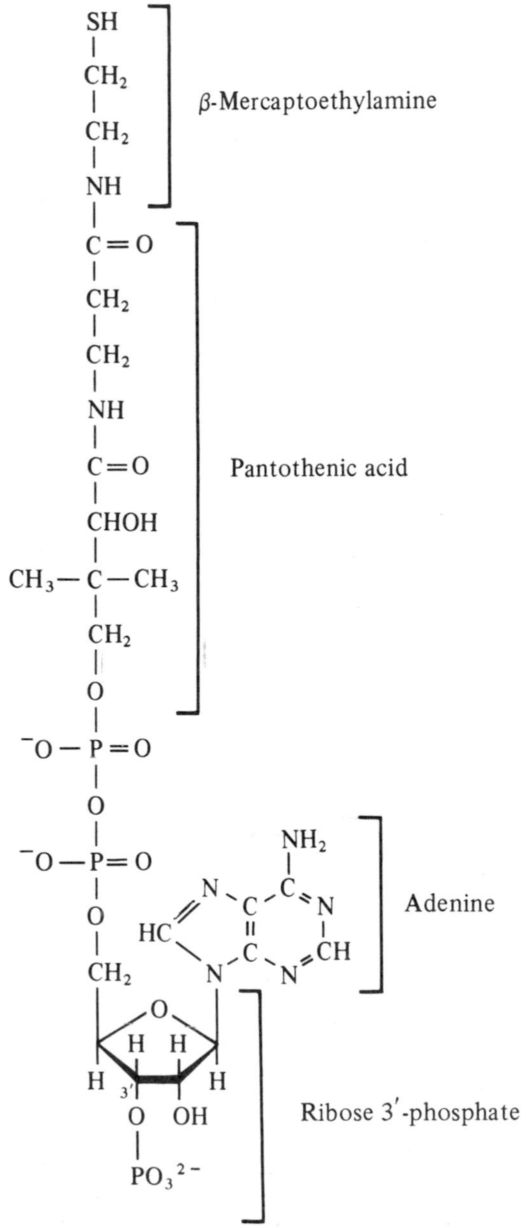

SH
|
CH₂
|
CH₂
|
NH } β-Mercaptoethylamine
|
C = O
|
CH₂
|
CH₂
|
NH
|
C = O } Pantothenic acid
|
CHOH
|
CH₃ — C — CH₃
|
CH₂
|
O
|
⁻O — P = O
|
O
|
⁻O — P = O
|
O
|
CH₂

Adenine

NH₂
N — C — N ... ring structure

Ribose 3′-phosphate

O OH
|
PO₃²⁻

Coenzyme A

FIGURE 15 Chemical structures of pantothenic acid and coenzyme A.

ABSORPTION AND METABOLISM

Pantothenic acid activity is widely distributed in feeds and foods, where it is present as a mixture of coenzyme forms. These forms of the vitamin are presumably hydrolyzed in the intestine. Serum contains predominantly free pantothenic acid. Cellular enzymes convert pantothenic acid to coenzyme A through a pathway that involves phosphorylation, addition of the β-mercaptoethylamine group, and, finally, addition of the nucleotide. Excess pantothenic acid is excreted in the urine, and changes in dietary intake can be followed by urinary excretion (Fox, 1984).

HYPERVITAMINOSIS

Pantothenic acid is generally regarded as nontoxic (Omaye, 1984). No adverse reactions have been reported in any species following the ingestion of elevated levels of pantothenic acid in the diet. Unna and Greslin (1941) determined an acute LD_{50} value for calcium pantothenate of about 1 g/kg of BW by parenteral injection for the rat but no toxicity at a dose of 10 g/kg of BW administered orally. They also reported that rats fed 200 mg of calcium pantothenate/day (about 20 g/kg of diet) for 190 days showed no adverse effects as far as growth nor any evidence of gross pathology. Wirtschafter and Walsh (1962) reported liver damage, as measured by lipid deposition and elevated serum glutamic/oxaloacetate transaminase activities, following the intramuscular injection of 20 mg of sodium pantothenate (about 80 mg/kg of BW) to rats. The severity of the response increased when greater doses were administered.

PRESUMED UPPER SAFE LEVELS

No adverse responses to the ingestion of pantothenic acid have been documented. The upper limits of presumed safe dietary levels cannot be established. It is clear, however, that dietary levels of at least 20 g of pantothenic acid/kg can be tolerated by most species.

SUMMARY

1. Pantothenic acid can be administered orally or in the diet at an intake of 10 g/kg of BW with no adverse effects.

2. Parenterally administered pantothenic acid has an acute LD_{50} of about 1 g/kg of BW for the rat. Doses of about 80 mg/kg of BW have been shown to be associated

with nonfatal liver damage. This amount is approximately 100 times the daily dietary requirement for the vitamin.

REFERENCES

Fox, H. M. 1984. Pantothenic acid. Pp. 437–458 in Handbook of Vitamins, L. J. Machlin, ed. New York: Marcel Dekker.

Lipmann, F., N. O. Kaplan, G. D. Novelli, L. C. Tuttle, and B. M. Guirard. 1947. Coenzyme for acetylation, a pantothenic acid derivative. J. Biol. Chem. 167:869.

Olson, R. E. 1984. Pantothenic acid. Pp. 377–382 in Present Knowledge in Nutrition, R. E. Olson, ed. Washington, D.C.: The Nutrition Foundation.

Omaye, S. T. 1984. Safety of megavitamin therapy. Adv. Exp. Med. Biol. 177:169.

Sauberlich, H. E. 1980. Pantothenic acid. Pp. 209–215 in Modern Nutrition in Health and Disease, R. S. Goodhart and M. E. Shils, eds. Philadelphia: Lea & Febiger.

Unna, K., and J. S. Greslin. 1941. Studies of the toxicity and pharmacology of pantothenic acid. J. Pharmacol. Exp. Ther. 73:85.

Williams, R. J. 1939. Pantothenic acid—A vitamin. Science 89:486.

Wirtschafter, Z. T., and J. R. Walsh. 1962. Hepatocellular lipid changes produced by pantothenic acid excess. Ann. Surg. 15:976.

Biotin

Biotin is distributed widely in low concentrations in plant and animal tissues. It was first described as the factor protecting against egg white injury caused by avidin, a glycoprotein that binds biotin in the intestinal tract and thus inhibits biotin absorption. Biotin is slightly soluble in water. This vitamin has a bicyclic structure consisting of a ureido ring fused with a tetrahydrothiophine ring bearing a valeric acid side chain (Figure 16).

NUTRITIONAL ROLE

Dietary Requirements of Various Species

Although animals have a metabolic requirement for biotin, it is not yet established that it is a dietary essential for all species, due to widespread intestinal synthesis. The administration of certain sulfa drugs can be used to induce biotin deficiencies. Dietary requirements have been established for poultry, but not for most other animals. The broiler chick needs 0.15 mg of biotin/kg of diet and the turkey poult, 0.2 mg/kg (National Research Council, 1984).

Eight stereoisomers of the biotin molecule are possible but only one, D-biotin (*cis*), is physiologically active. Much of the biotin in natural materials is present in the bound form, of ϵ-N-biotinyl-L-lysine in proteins. The biological availability of this form, which is called biocytin, depends upon the digestibility of the proteins in which it is found. The supplementary form generally used for animal feeds is synthetic D-biotin.

Biochemical Functions

Biotin is known to take part in metabolic carboxylation reactions (Bonjour, 1984). Most biotin-dependent reactions in mammals are energy-dependent. The reaction involves the cleavage of adenosine triphosphate (ATP) to adenosine diphosphate (ADP) and inorganic phosphate. Most biotin-dependent reactions in mammalian tissues appear to be of this type. The exception is a transcarboxylation that is not energy-dependent. The biotin-catalyzed carboxylase, systems consist of three types of subunits: biotin carboxylase, carboxyl transferase, and carboxyl carrier protein. Biotin is covalently linked to the carboxyl carrier protein through a peptide bond to the ϵ-amino group of lysine (that is, as biocytin). The most important of the biotin-dependent carboxylation enzymes are pyruvate carboxylase, acetyl CoA carboxylase, and propionyl CoA carboxylase.

ABSORPTION AND METABOLISM

Biotin appears to be absorbed well from the small intestine (Marks, 1979), although the protein-bound forms in feeds are not readily available to animals. Present evidence suggests that the vitamin is not retained well. A high proportion of administered biotin has been recovered intact in the urine of rats and humans following parenteral adminstration (Fraenkel-Conrat and Fraenkel-Conrat, 1952; Wright et al., 1956). Biotin excretion, like that of most water-soluble vitamins, is closely related to intake.

Biotin acts with a number of carboxylases, which results in the movement of carbon atoms between cellular compartments (Murthy and Mistry, 1977). The biotin-dependent enzyme pyruvate carboxylase converts pyruvate to oxaloacetic acid. This reaction is important in gluconeogenesis, the formation of glycerol, and the synthesis of acetylcholine. Two other biotin enzymes, propionyl-CoA carboxylase and acetyl-CoA carboxylase, are involved in the synthesis of succinyl CoA.

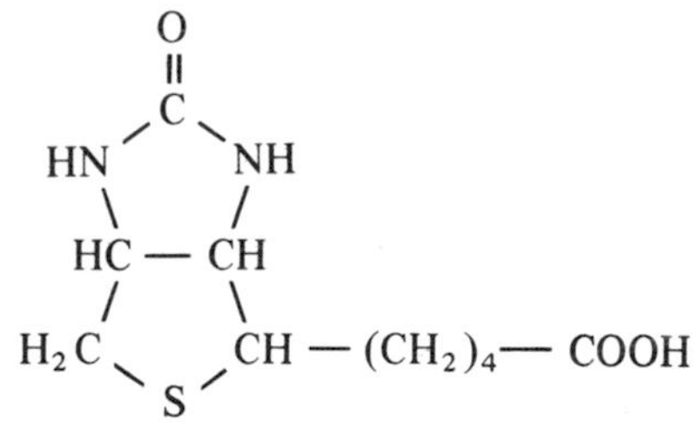

O
‖
C
HN   NH

HC — CH

H₂C   CH — (CH₂)₄— COOH
   S

Biotin

FIGURE 16   Chemical structure of biotin.

The former is of major importance in energy utilization in ruminants. Biotin is also required in the fixation of 6-carbon in purines. Thus, it is important in RNA and DNA syntheses.

HYPERVITAMINOSIS

The effects of the administration of high levels of biotin are summarized in Table 15. Very limited data are available.

Comben (1979) reported the results of supplementing the diet of breeding sows with 2 mg of biotin/kg of diet for a period of 4 months. He noted reduction in lameness within 5 to 6 weeks, as well as an improvement in reproduction. No adverse effects were recorded.

Bryant et al. (1985a,b,c) conducted a series of trials in which female swine were fed supplemental biotin at levels up to 0.44 mg/kg of diet. Growing animals given a supplementary level of 0.22 mg of biotin/kg of diet showed improved foot health. Growth was unaffected. No adverse effects were noted. During the subsequent breeding period, which extended over four reproductive cycles, the sows were fed either no supplemental biotin or 0.44 mg/kg of diet. Improved foot health, hair coat, and reproduction were noted. No adverse effects were recorded.

Adams et al. (1967) fed a diet based on corn, milo, and soybean meal, which contained 0.29 or 0.4 mg of biotin/kg, to early-weaned pigs. The researchers reported improved growth and feed efficiency over a 122-day growth period. No adverse effects of the higher level of biotin were recorded. Brooks et al. (1977) increased the biotin level of breeding sows from 0.15 mg/kg of diet to 0.35 or 0.4 mg/kg. They observed a 28 percent reduction in the incidence of foot lesions. Litter productivity was also increased. No adverse effects of the higher biotin levels were observed.

Arends et al. (1971) fed 0.15, 0.25, or 0.33 mg of biotin/kg of diet to 32-week-old breeding turkey hens for 12 or 14 weeks. No significant effects on egg production or hatchability were noted. However, the egg biotin content was raised from 0.13 mg/kg of egg weight with the control diet to 0.25 mg/kg with the diet containing 0.33 mg of biotin/kg. The mortality of poults from hens fed the biotin-supplemented diets was reduced. The progeny were fed up to 0.37 mg of biotin/kg of diet until they were 4 weeks old. Liver biotin concentration increased linearly with the dietary biotin level. Poults fed the highest level of biotin had hepatic biotin concentrations of 0.91 μg/g.

Brewer and Edwards (1972) fed diets containing 0.02, 0.04, 0.06, 0.1, 0.18, or 0.34 mg of biotin/kg of diet to breeding broiler hens for a 10-week period. Egg production and fertility were improved with diets containing up to 0.1 mg of biotin/kg. Egg biotin concentration increased linearly with increasing levels of the vitamin in the diet. A biotin level of 0.1 mg/kg of diet produced 0.2 μg/g of yolk weight; a biotin level of 0.34 mg/kg of diet produced 0.6 μg/g of yolk weight. No adverse effects of the higher biotin levels were noted.

Whitehead and Randall (1982) added 0.04 to 0.5 mg of biotin/kg to the diets of growing broiler chickens from 1 day to 8 weeks of age. They reported no adverse effects on mortality.

Paul et al. (1973a,b) administered to unmated 3- to 4-month-old female rats 50 mg of biotin/kg of BW in two subcutaneous injections of biotin dissolved in 0.5 ml of 0.1 N sodium hydroxide (NaOH). The rats were bred 7, 14, or 21 days after treatment. They were sacrificed 15 or 22 days after gestation. The biotin-treated rats showed irregular estrous cycles. Atrophic changes were observed in the corpora lutea. A large number of the treated animals resorbed their fetuses by the twenty-second day of gestation, even when mated within 7 days following treatment. Fetal and placental weights were also reduced.

Paul and Duttagupta (1975) administered a subcutaneous dose of 100 mg of biotin/100 g of BW to female rats on the first and second days of gestation. They reported resorption of fetuses and placentas in 7 of the 8 treated animals by 21 days of gestation. In a subsequent investigation, Paul and Duttagupta (1976) administered a subcutaneous dose of D-biotin in 0.1 N NaOH to female rats on the fourteenth and fifteenth days of gestation. Resorption of fetuses and placentas occurred in 2 out of 11 animals. Maternal, uterine, fetal, and placental weights were reduced in the others.

Mittelholzer (1976) was unable to reproduce those effects using a similar protocol. In this study, pure D-biotin was administered to 190- to 240-g female rats that were known to be cycling. On the day of vaginal estrus, each rat, with the exception of those in the control group, was given subcutaneous injections totalling 5 or 50 mg of

72

TABLE 15 Research Findings of High Levels of Biotin in Animals

Species and No. of Animal	Age or Weight	Administration Form	Amount	Duration	Route	Effect	Reference
Birds							
Chickens, broilers, males, 1,260	1 d		0, 0.04, 0.08, 0.1, 0.13, 0.2, 0.5 mg/kg BW	8 wk	Diet	No adverse effects on mortality	Whitehead and Randall, 1982
Chickens, broilers, 126	Breeder hens	Biotin	0.02–0.32 mg/kg diet	10 wk	Diet	Egg production and fertility improved up to 0.1 mg/kg	Brewer and Edwards, 1972
Turkeys, medium whites or broad whites, 1,333	Breeder hens 32 wk	Biotin	0.15–0.33 mg/kg diet	12 or 14 wk	Diet	No effects on egg production or hatchability; egg biotin content raised from 0.13 to 0.25 mg/kg	Arends et al., 1971
Turkeys, medium whites or broad whites, 120	Hatched poults	Biotin	0.09–0.36 mg/kg diet	4 wk	Diet	No effects on growth; liver biotin increased linearly to 0.91 mg/kg with highest level	Arends et al., 1971
Swine, sows, 240	Mature	D-biotin	2 mg/kg diet	4 mo	Diet	Reduction in lameness; no adverse effects noted	Comben, 1979
Swine, sows, 20	Mature, 2nd and 3rd parity	D-biotin	0.35–0.4 mg/kg diet	Gestation and lactation	Diet	Fewer foot lesions and better litter productivity than with 0.15 mg/kg	Brooks et al., 1977
Swine, females, 116	100 kg	D-biotin	0.44 mg/kg	Over four reproductive cycles	Diet	Improved reproduction, foot health, and hair coat; no adverse effects noted	Bryant et al., 1985a, b
Swine, 96	7 kg	D-biotin	0.4 mg/kg diet	122 d	Diet	Growth and feed efficiency better than with 0.29 mg/kg	Adams et al., 1967
Swine, females	28.6 d (6.7 kg BW)	D-biotin	0.22 mg/kg diet (supplemental)	Up to 154 d	Diet	Improved foot health; no effects on growth; no adverse effects noted	Bryant et al., 1985c
Rats, females, 36	3 to 4 mo	Biotin	50 mg/kg BW	In two doses within 24 h	SC	Resorption of fetuses noted at 14 d and 21 d; fetal number and number of implantation sites per dam reduced; fetal and placental weights reduced	Paul et al., 1973a
Rats, females, 18	3 mo	Biotin	50 mg/kg BW	In two doses within 24 h	SC	Subsequent irregularity of estrous cycle; atrophic changes noted in corpora lutea	Paul et al., 1973b
Rats, females, 8	Gestation	D-biotin (0.1 N NaOH)	100 mg/kg BW	1st and 2nd successive d of pregnancy	SC	Resorption of fetuses and placentas in 7 by 21 d	Paul and Duttagupta, 1975
Rats, females, 11	3.5 mo	D-biotin (0.1 N NaOH)	100 mg/kg BW	14th and 15th successive d of pregnancy	SC	Resorption of fetuses and placentas in two rats; maternal, uterine, fetal and placental weights reduced	Paul and Duttagupta, 1976
Rats, females, 72	190–240 g	D-biotin	5 or 50 mg/kg BW	In two doses within 24 h	SC	No significant reduction in breeding success	Mittelholzer, 1976

biotin/kg of BW with 0.1 N NaOH as the carrier vehicle. Treatment consisted of two doses spaced 6 hours apart. It was found that 88 percent of rats receiving 5 mg/kg carried fetuses at 21 days of gestation. This rate was 71 percent in the high-biotin group and 67 percent in the control group. No significant differences were noted in implantation rate, number of resorption sites, fetal weight, placental weight, or ovarian weight.

PRESUMED UPPER SAFE LEVELS

Insufficient data are available to support estimates of the maximum dietary tolerable levels of biotin. Results from one laboratory suggest that biotin may be toxic to fetal rats when it is administered subcutaneously to the dam. This effect, however, was not reproduced by another researcher who followed a similar protocol. Studies with poultry and swine indicate that these species can safely tolerate dietary levels of at least 4 times their nutritional requirements of biotin. In view of the poor retention of biotin, it is probable that these species can tolerate much higher levels.

SUMMARY

1. Biotin is a water-soluble vitamin that many species of animals require in the diet. For other species, gut microbial synthesis provides sufficient biotin. The vitamin appears to be well absorbed from the gut, but is not well retained.

2. Studies with poultry and swine indicate that these species can safely tolerate dietary levels of 4 to 10 times their nutritional requirements of biotin. Because this vitamin is not well retained, the maximum tolerable level of biotin may be much higher.

REFERENCES

Adams, C. R., C. E. Richardson, and T. J. Cunha. 1967. Supplemental biotin and vitamin B₆ for swine. J. Anim. Sci. 26:903. (Abstr.)

Arends, L. G., E. W. Kienholz, J. V. Schutze, and D. D. Taylor. 1971. Effect of supplemental biotin on reproductive performance of turkey breeder hens and its effect on the subsequent progeny's performance. Poult. Sci. 50:208.

Bonjour, J. P. 1984. Biotin. P. 403 in Handbook of Vitamins, L. J. Machlin, ed. New York: Marcel Dekker.

Brewer, L. E., and H. M. Edwards, Jr. 1972. Studies on the biotin requirement of broiler breeders. Poult. Sci. 51:619.

Brooks, P. H., D. A. Smith, and V. C. R. Irwin. 1977. Biotin—Supplementation of diets, the incidence of foot lesions, and the reproductive performance of sows. Vet. Rec. 101:46.

Bryant, K. L., E. T. Kornegay, J. W. Knight, K. E. Webb, Jr., and D. R. Notter. 1985a. Supplemental biotin for swine. II. Influence of supplementation to corn-and-wheat-based diets on reproductive performance and various biochemical criteria of sows during four parities. J. Anim. Sci. 60:145.

Bryant, K. L., E. T. Kornegay, J. W. Knight, H. P. Veit, and D. R. Notter. 1985b. Supplemental biotin for swine. III. Influence of supplementation to corn-and-wheat based diets on the incidence and severity of toe lesions, hair and skin characteristics and structural soundness of sows housed in confinement during four parities. J. Anim. Sci. 60:154.

Bryant, K. L., E. T. Kornegay, J. W. Knight, K. E. Webb, Jr., and D. R. Notter. 1985c. Supplemental biotin for swine. I. Influence on feedlot performance, plasma biotin and toe lesions in developing gilts. J. Anim. Sci. 60:136.

Comben, N. 1979. Biotin for Breeding Sows. Field Experiences 1976–1978. Publication #1979-270-79-955. Basel, Switzerland: F. Hoffman-LaRoche.

Fraenkel-Conrat, J., and H. Fraenkel-Conrat. 1952. Metabolic fate of biotin and of avidin-biotin complex upon parenteral administration. Biochem. Biophys. Acta 8:66.

Marks, J. 1979. A Guide to Vitamins. Lancaster, England: Medical and Technical Publishing.

Mittelholzer, E. 1976. Absence of influence of high doses of biotin on reproductive performance in female rats. Int. J. Vit. Res. 46:33.

Murthy, P. N. A., and S. P. Mistry. 1977. Biotin. Prog. Food Nutr. Sci. 2:405.

National Research Council. 1984. Nutrient Requirements of Poultry. 8th rev. ed. Washington, D.C.: National Academy Press.

Paul, P. K., and P. N. Duttagupta. 1975. The effect of an acute dose of biotin at the pre-implantation stage and its relation with female sex steroids in the rat. J. Nutr. Sci. Vitaminol. 21:89.

Paul, P. K., and P. N. Duttagupta. 1976. The effect of an acute dose of biotin at the post-implantation stage and its relation with the female sex steroids in the rat. J. Nutr. Sci. Vitaminol. 22:181.

Paul, P. K., P. N. Duttagupta, and H. C. Agarwal. 1973a. Antifertility effect of biotin and its amelioration by estrogen in the female rat. Curr. Sci. 42:613.

Paul, P. K., P. N. Duttagupta, and H. C. Agarwal. 1973b. Effects of an acute dose of biotin on the reproductive organs of the female rat. Curr. Sci. 42:206.

Whitehead, C. C., and C. J. Randall. 1982. Interrelationships between biotin, choline and other B-vitamins and the occurrence of fatty liver and kidney syndrome and sudden death syndrome in broiler chickens. Br. J. Nutr. 48:177.

Wright, L. D., E. L. Cresson, and C. A. Driscoll. 1956. Biotin derivatives in human urine. Proc. Soc. Exp. Biol. Med. 91:248.

Vitamin B$_{12}$

Vitamin B$_{12}$ (cobalamin) is unique among vitamins in that it is synthesized in nature only by microorganisms. It is the last vitamin to have been discovered (in the late 1940s) and is the most potent on a weight basis. Vitamin B$_{12}$ deficiencies are characterized by a wide variety of signs in various animal species. The natural concentrations of this vitamin in feeds are generally low. A synthetic form is commonly used as a feed supplement.

NUTRITIONAL ROLE

Dietary Requirements of Various Species

All nonruminant species require dietary sources of vitamin B$_{12}$. The required amounts are low because of the presence of microbial sources of the vitamin in the environment (e.g., manure) and bacterial synthesis in the gastrointestinal tract. However, the latter contribution may be of questionable significance. Supplementation of diets based solely on plant feedstuffs is essential. Vitamin B$_{12}$-deficient swine may show macrocytic hyperchromic anemia, neuropathies, reproductive failures, and dermatitis; chickens may show abnormal feathering; and rats may show porphyrin-caked whiskers. The estimated requirements of most species for vitamin B$_{12}$ range from 9 to 22 μg/kg of diet.

Biochemical Functions

Vitamin B$_{12}$ normally occurs in feeds bound to protein in the methyl or 5'-deoxyadenosyl forms, each of which is known to be a coenzyme in only a single reaction in animal metabolism. The methyl form (methyl cobalamin) is required as a carrier of the methyl group from N^5-methyltetrahydrofolate to homocysteine in the conversion of the latter to methionine. The 5'-deoxyadenosyl form (adenosylcobalamin) is required in the conversion of methylmalonyl CoA to succinyl CoA, an important step in the metabolism of propionic acid.

FORMS OF THE VITAMIN

The structure of vitamin B$_{12}$ is shown in Figure 17. It consists of a corrin ring system with a central cobalt atom. Cyanocobalamin is the usual form of the vitamin used in supplementing animal feeds. It contains a cyanide group as an artifact of the preparation process attached to the central cobalt atom. Little, if any, of this form is believed to occur naturally. However, other forms of the vitamin in which cyanide is replaced by another group occur naturally. These include hydroxycobalamin that has been isolated from liver and nitritocobalamin that has been isolated from microorganisms. Other forms, which are found commonly in feeds, are methylcobalamin (the cyanide group replaced with a methyl group) and 5'-deoxyadenosylcobalamin (the cyanide group replaced with a deoxyadenosylcobalamin group). All of the above-mentioned forms have vitamin B$_{12}$ activity.

ABSORPTION AND METABOLISM

Vitamin B$_{12}$ is synthesized by the intestinal microflora in nonruminant species and by rumen microbes in ruminants. This source is normally sufficient to meet the needs of ruminants. It is not known how much of the source can be absorbed in nonruminants, however.

Absorption of this water-soluble vitamin is mainly or exclusively in the ileum and is facilitated by the presence of an intrinsic factor released in gastric juice. Failure to produce the intrinsic factor (for example, as the result of pernicious anemia or total gastrectomy) results in failure to absorb vitamin B$_{12}$. Denker (1983) injected pregnant mice with radiolabeled vitamin B$_{12}$ (cyanoco-

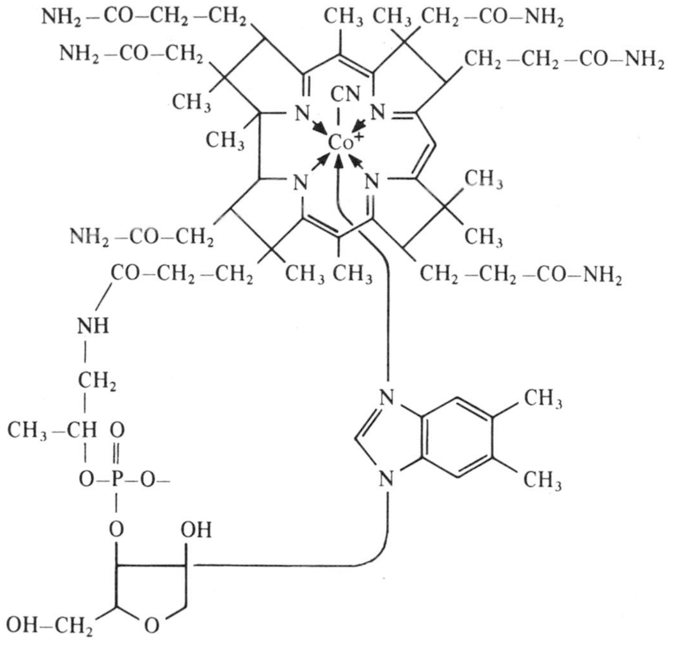

Vitamin B$_{1\,2}$

FIGURE 17   Chemical structure of the cyano form of vitamin B$_{12}$ (cyanocobalamin).

balamin-^{57}Co) intravenously at a dose rate of 3.2 mg/kg of BW. Three hours later, he recovered 79 percent of the vitamin in the placenta, 1.5 percent in the serum, 2.1 percent in the fetus, 2.3 percent in the liver, and 15 percent in the kidney. With continued intake of the vitamin, animals show tissue storage of cobalamins principally in the liver (30 to 60 percent of the total body load), but also at lower levels in the kidney, heart, spleen, and brain (Ellenbogen, 1984). In humans, the tissue storage

of the vitamin is so great that signs of vitamin B$_{12}$ deficiency may not appear for months or years after the vitamin has stopped being excreted in urine and bile.

Vitamin B$_{12}$ acts in a number of roles that are important in the functioning of tetrahydrofolate, which is the facilitation of folate entry into cells and the transfer of the methyl group from methyltetrahydrofolate to homocysteine. In B$_{12}$ deficiency, tetrahydrofolate is thought to accumulate as methyltetrahydrofolate, which is unable to transfer methyl groups in the synthesis of thymidine. The resulting defect in DNA synthesis, which is characteristic of folate deficiency, also is produced by a B$_{12}$ deficiency. Because vitamin B$_{12}$ is required in the conversion of propionate to succinate, deficient animals excrete methylmalonic acid in the urine.

The criteria for vitamin B$_{12}$ adequacy include normal rates of growth, hematopoiesis, reproduction, offspring viability, and liver concentrations of the vitamin.

HYPERVITAMINOSIS

A summary of the effects of vitamin B$_{12}$ administration in animals is shown in Table 16. Schaefer et al. (1949) fed 15 or 30 μg of vitamin B$_{12}$/kg of diet to day-old leghorn chickens to 4 weeks of age and found no adverse effects of the higher level. Traina (1950) administered to mice intraperitoneal doses of 7.5, 15, and 30 μg of vitamin B$_{12}$, or a subcutaneous dose of 30 μg. Traina observed signs of toxicity with doses of 15 μg (1.36 mg/kg of BW) and higher. Winter and Mushett (1951) administered doses of up to 1,600 mg of vitamin B$_{12}$/kg of BW to mice by either the intraperitoneal or intravenous route and reported no adverse effects on growth or survival.

TABLE 16   Research Findings of High Levels of Vitamin B$_{12}$ in Animals

Species and No. of Animal	Age or Weight	Administration				
		Amount	Duration	Route	Effect	Reference
Chickens, leghorns, 11–60/group	1 d	15 or 30 μg/kg diet	4 wk	Diet	Growth rate similar; no adverse effects	Schaefer et al., 1949
Mice, 2	Gestating 36 g	0.114 μg	2 injections 10 min apart	IV	Concentration greatest in placenta	Denker, 1983
Mice, albinos, 10/group	11 g	7.5, 15, or 30 μg	Single dose	IP	7.5 μg, no adverse effects; 15 μg, 20% mortality; 30 μg, 100% mortality	Traina, 1950
Mice, albinos, 10/group	11 g	30 μg	Single dose	SC	100% Mortality	Traina, 1950
Mice, 3–10/group	20 g	100–1,600 mg/kg BW	Single dose	IP	No mortality or adverse effects on growth	Winter and Mushett, 1951
Mice, 3/group	20 g	800 and 1,600 mg/kg BW	Single dose	IV	No mortality or adverse effects on growth	Winter and Mushett, 1951

NOTE: The form was B$_{12}$ in all cases.

They suggested that the effects found by Traina (1950) may have been due to the presence of toxic impurities in the sample of vitamin used.

PRESUMED UPPER SAFE LEVELS

Insufficient data are available to support estimates of the maximum dietary tolerable levels of vitamin B_{12}. Data from a single chick study suggest that 3 times the vitamin B_{12} requirement of that species can be included safely in the diet, however. Mouse data suggest that dietary levels of at least several hundred times the requirement are safe.

SUMMARY

1. Vitamin B_{12} is a water-soluble vitamin that is stored principally in the liver. It is required in the diets of non-ruminant animals.

2. Data from mouse studies suggest that vitamin B_{12} is innocuous when administered intraperitioneally or intravenously in relatively high doses and that dietary levels of at least several hundred times the requirement are safe.

REFERENCES

Denker, L. 1983. Placental accumulation of [57]Co-vitamin B_{12} in mice studied by light and electron-microscopic autoradiography. Placenta 4:207.

Ellenbogen, L. 1984. Vitamin B_{12}. P. 497 in Handbook of Vitamins, L. J. Machlin, ed. New York: Marcel Dekker.

Schaefer, A. E., W. D. Salmon, and D. R. Strength. 1949. Interrelationship of vitamin B_{12} and choline. II. Effect on growth of the chick. Proc. Soc. Exp. Biol. Med. 71:202.

Traina, V. 1950. Toxicity studies on vitamin B_{12} in albino mice. Arch. Pathol. 49:278.

Winter, C. A., and C. W. Mushett. 1951. Absence of toxic effects from single injections of crystalline vitamin B_{12}. J. Am. Pharm. Assoc. 39:360.

Choline

Although choline is not a true vitamin in the classical sense, it is an important nutrient as a source of labile methyl groups. It is synthesized by many animal species. Certain physiological states or clinical disorders may result in relative deficiencies and subsequent needs for choline dietary supplements, however. Choline was originally isolated from hog bile.

NUTRITIONAL ROLE

Dietary Requirements of Various Species

Dietary requirements for choline have been established for the young of several species including the chicken, pig, rat, and dog (see the appendix table). Most animals can produce all of the choline they need by hepatic synthesis. This synthesis can be insufficient for the needs of rapidly growing poultry or for the young of other species when fed diets deficient in methyl groups, however. In these cases, dietary supplements of choline are required to alleviate growth depression and/or hepatic steatosis. Choline-deficient chicks and poults also show a condition called chondrodystrophy (perosis). Chondrodystrophy is characterized by hemorrhaging and flattening of the tibiometatarsal joint, rotation of the metatarsus, and, ultimately, displacement of the Achilles tendon from its chondyles, resulting in crippling (National Research Council, 1984). The chick's choline requirement is substantial (aproximately 1,300 mg/kg of diet) until about 13 weeks of age, after which time endogenous synthesis can apparently satisfy physiological demands.

Berry et al. (1943), Marvel et al. (1943), and Mishler et al. (1946) found that the apparent choline requirements of chicks are increased by including soybean meal in the diet. Soybeans contain significant amounts (in excess of 2,500 mg/kg) of choline. Therefore, it has been suggested that the biological availability of choline in this feedstuff may be relatively poor. Molitoris and Baker (1976) reported that choline in soybean meal is only 60 to 75 percent available for supporting chick growth. The chick growth data of Fritz et al. (1967) presents a somewhat higher estimate of 85 to 89 percent.

Biochemical Functions

Choline has three important functions in metabolism (Chan, 1984). As the acetyl ester, acetylcholine, it serves as a neurotransmitter. It is also metabolized to phosphatidyl choline (lecithin). In this form, choline has structural functions in biological membranes and in tissue lipid utilization. Choline is also oxidized to betaine, serving as a source of labile methyl groups for the formation of methionine from homocysteine and of creatine from guanidoacetic acid.

FORMS OF THE VITAMIN

Choline occurs in biological tissues in the free form (trimethylethanolamine) and as a component of lecithin, acetylcholine, and other phospholipids (see Figure 18). The form most commonly used for supplementation of diets is choline hydrochloride, which may be in liquid, deliquescent, or solid form. Various investigators have also used choline dihydrogen citrate and cytidine diphosphate choline (CDP-choline).

ABSORPTION AND METABOLISM

Little is known of the absorption of choline. Because intestinal microflora break down choline in the large

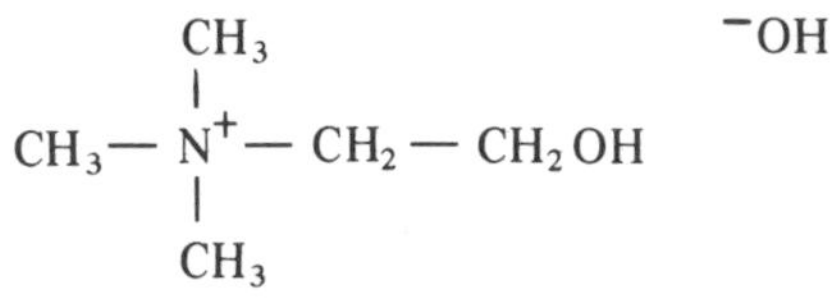

$$CH_3 - \overset{\overset{\displaystyle CH_3}{|}}{\underset{\underset{\displaystyle CH_3}{|}}{N^+}} - CH_2 - CH_2OH \qquad {}^-OH$$

Choline
(β-hydroxyethyltrimethyl-
ammonium hydroxide)

FIGURE 18   Chemical structure of choline.

intestine to trimethylamine, absorption can be assumed to take place in the small intestine.

All species are capable of synthesizing choline in the liver by the methylation of ethanolamine, which uses methyl groups from *S*-adenosyl methionine. The process occurs in two steps, each involving a different methyl transferase. Limitations in the availability of methyl groups, therefore, can reduce endogenous choline production. Methionine has a choline-sparing effect. For swine, it has been estimated that when methionine is fed at levels in excess of that required for normal rates of protein synthesis, 4.3 mg of methionine was equal to 1 mg of choline in providing methylating capacity (National Research Council, 1979).

According to Mookerjea (1971), the hepatic accumulation of lipids in both choline- and methyl group-deficient rats is due to decreased formation of low-density lipoproteins, which results from inadequate amounts of lecithin. Choline is important in the transformation of the immunoreactive cells. Nauss and Newberne (1980) reported that the thymus gland was hypocellular in rats born from choline-deficient parents. This symptom suggests that defective cellular proliferation is due to impaired DNA synthesis. Beisel (1982) reported impaired chemotaxis in macrophages and depressed T-lymphocyte response.

Plasma choline may be used as a criterion of adequacy. Growth rate, satisfactory reproduction, and lipid levels of livers and kidneys also have been used.

HYPERVITAMINOSIS

Neumann et al. (1949), Kroenig and Pond (1967), and Dobson (1971) have reported no adverse effects of choline supplementation in swine. In fact, the growth rate improved (see Table 17).

Studies with poultry indicated similar tolerance to high levels of choline. Ketola and Nesheim (1974) provided levels of 500 to 2,500 mg of choline (as choline dihydrogen citrate)/kg of purified diet to day-old leghorn-type of chickens. They noted no adverse effects

on growth over a 21-day period. In fact, they noted that increasing the level of the vitamin above 500 mg/kg of diet reduced chondrodystrophy to zero. Some leghorn-type of laying hens at 30 weeks of age and 80 percent of egg production were fed purified diets containing either no choline chloride or 1,400 mg/kg. Ketola and Nesheim noted no adverse effects on egg production during a 12-week period. Supplementation of the diet with choline chloride led to significant increases in both BW gain and feed intake. Jukes (1941) fed diets containing either no choline or 2,000 mg/kg (presumed to be the chloride form) to turkey poults from hatching. Jukes reported an improvement in growth and a reduction in chondrodystrophy. Crawford et al. (1969) and March and MacMillan (1979) reported no deleterious effects in laying hens fed diets containing up to 5,730 mg choline/kg (added as choline HCl).

Other studies have revealed adverse effects of high levels of choline. Saville et al. (1967) fed day-old broilers graded levels of choline chloride from 400 to 2,200 mg/kg of diet. They noted hyperexcitability and muscular incoordination after 7 weeks in those animals fed the 2,200 mg/kg. Similar signs developed later in the other treatment groups. Growth after 6 weeks of age was depressed in the groups receiving 1,320, 1,760, and 2,200 mg/kg. The problem was overcome by withdrawal of the choline chloride or by provision of additional pyridoxine. Deeb and Thornton (1959) fed semipurified diets containing supplementary choline chloride at levels up to 8,800 mg/kg to day-old broiler chickens to 4 weeks of age. Growth rate was maximized with 880 to 1,760 mg choline chloride/kg of diet. They reported a depression in BW and feed efficiency with a dietary level exceeding 2,200 mg/kg. This level was slightly above the one Jukes (1941) found to be well tolerated by broilers.

Davis (1944a) showed that daily oral administration of 5 g of soybean lecithin (equivalent to 150 mg of choline) to dogs resulted in a maximum number of erythrocyte reduction that took place after 12–25 days. This condition persisted for at least 10 days after the cessation of lecithin feeding. Similarly, daily doses of 8 mg of choline hydrochloride/kg of BW also significantly reduced red blood cell numbers. In this case, more than 10 days were required for maximum red blood cell number depression to occur. Davis (1944a) suggested that choline administration depressed erythropoiesis by increasing the oxygen supply to the bone marrow.

Davis (1944b) found that choline chloride induced a hyperchromic anemia in about 15 dogs. The anemia was produced by giving the dogs single doses of 10 mg/kg/day of choline chloride by stomach tube. Once the anemia was established, the same dose was continued twice daily. When the erythrocyte numbers were further reduced, a third daily dose was added. One dog was placed

TABLE 17 Research Findings of High Levels of Choline in Animals

Species and No. of Animal	Age or Weight	Administration Amount	Form	Duration	Route	Effect	Reference
Birds							
Chickens, New Hampshire X Delawares	1 d	0–8,800 mg/kg diet	Choline chloride	4 wk	Diet	BW and feed efficiency reduced slightly with more than 2,200 mg/kg	Deeb and Thornton, 1959
Chickens, leghorns 830	Laying	1,340 mg/kg diet	Choline HCl	28–48 wk	Diet	Egg production, mortality unaffected; reduction in liver lipid of 23.5–28.2%	March, 1981
Chickens, female leghorns, 15/group	Laying (80%); 30 wk old	1,400 mg/kg diet	Choline chloride	12 wk	Diet	No adverse effects on egg production; increased feed intake and weight gain	Ketola and Nesheim, 1974
Chickens, male leghorns, 24/group	1 d	500–2,500 mg/kg diet	Choline dihydrogen citrate	3 wk	Diet	No adverse effects on growth	Ketola and Nesheim, 1974
Chickens, broilers, 200/group	1 d	Supplement of 440–2,200 mg/kg diet	Choline chloride	10 wk	Diet	Signs of pyridoxine deficiency in group receiving the highest level at 7 wk, later in other groups. Growth from 6 wk depressed in groups receiving 1,320; 1,760; and 2,200 mg/kg choline HCl; problem overcome with added pyridoxine	Saville et al., 1967
Chickens, 185	Laying	Supplement of 3,834–5,228 mg/kg diet	Choline HCl	Up to 252 d	Diet	Some eggs with fishy taint; no other adverse effects	March and MacMillan, 1979
Chickens, 60	Laying	Supplement of 4,600–7,330 mg/kg diet	Choline HCl	8 mo	Diet	Decreased liver lipid; feed intake increased with 7,330 mg/kg; no adverse effects	Crawford et al., 1969
Turkeys, 24–30	1 d	2,000 mg/kg diet	Choline HCl[a]	28 d	Diet	Growth improved and incidence of perosis reduced; no adverse effects	Jukes, 1941
Dogs, 2	Mature	10 mg/kg BW 3/d	Choline HCl	90 d	Stomach tube	Erythrocyte numbers reduced	Davis, 1944b
Dogs, 1	Mature	10 mg/kg BW/d	Choline HCl	d 7	Stomach tube	Erythrocyte numbers reduced	Davis, 1944c
Dogs, 1	Mature	10 mg/kg BW 2/d	Choline HCl	d 19	Stomach tube	Erythrocyte numbers reduced	Davis, 1944b
Dogs, 1	Mature	10 mg/kg BW 3/d	Choline HCl	d 60	Stomach tube	Erythrocyte numbers reduced	Davis, 1944b
Dogs, 4	Mature	10 mg/kg BW/d	Choline HCl	25 d + until anemia established	Stomach tube	Erythrocyte numbers reduced; in 5 dogs, administration of 3 daily doses produced reductions from 30–43%	Davis, 1944b
Dogs, 4	Mature	10 mg/kg BW 2/d	Choline HCl	Until red cell level lowered	Stomach tube	Erythrocyte numbers reduced; in 5 dogs administration of 3 daily doses produced reductions from 30 to 43%	Davis, 1944b

TABLE 17—*Continued*

Species and No. of Animal	Age or Weight	Administration Amount	Form	Duration	Route	Effect	Reference
Dogs, 4	Mature	10 mg/kg BW 3/d	Choline HCl	After red cell levels lowered	Stomach tube	3 daily doses produced reductions from 30 to 43%	Davis, 1944b
Dogs, 6	1.8 kg	Supplement of 1,500 mg/kg diet	Choline HCl	32–57 d	Diet	Improvement in growth; no adverse effects	McKibbin et al., 1944
Dogs, 4	1.5–2.7 kg	Supplement of 2,000 mg/kg diet	Choline HCl	10–50 d	Diet	Improvement in growth and liver function and reduction in liver lipid content; no adverse effects reported	McKibbin et al., 1945
Dogs, 4	Mature	5 g/d	Soybean lecithin 3% choline	80 d	Diet	After latent period of at least 5 d, erythrocyte numbers gradually reduced; max diminutions of 15–20% reached after 12–25 d of lecithin feeding	Davis, 1944a
Dogs, 2	Mature	8 mg/kg BW/d 0.5 mg/kg BW (for 18 d)	Choline HCl administered with atropine	35 d	Stomach tube	Required additional 10 d after atropine cessation to show comparable depressions	Davis, 1944a
Dogs, 2	Mature	8 mg/kg BW/d	Choline HCl	35 d	Stomach tube	After 15 d, erythrocyte numbers significantly reduced	Davis, 1944a
Humans with movement disorders, 8	Mature	Up to 20 g/d	Choline HCl	4 wk	Oral	16–20 g associated with peak plasma concentration; rapid disappearance after dosing; some clinical improvement	Hollister et al., 1978
Humans with movement disorders, 8	Mature	5 g/d	Choline HCl	Single dose	Oral	Plasma concentration peak at 4 h	Hollister et al., 1978
Mice, male and female albinos, 106	18–26 g	5–10 mg	Choline HCl	Single injection	IP	LD_{50}, 320 mg/kg BW	Hodge and Goldstein, 1942
Mice, male and female, Swiss CD-1 holuxenic	18–22 g	Varied	CDP-choline	Single dose	IV	LD_{50}, 4,600 mg/kg BW LD_0, 3,500 mg/kg BW	Agut et al., 1983
Mice, male, Swiss CD-1 holuxenic	18–22 g	Varied	Choline HCl	Single dose	IV	LD_{50}, 53 mg/kg LD_0, 21.5 mg/kg	Agut et al., 1983

Animal	Age/weight	Dose	Compound	Duration	Route	Effects	Reference
Mice, male and female, Swiss CD-1 holuxenic	18–22 g	Varied	CDP-choline	Single dose	Oral	LD_{50}, indeterminate LD_0, 14,000 mg/kg	Agu et al., 1983
Mice, male and female, Swiss CD-1 holuxenic	18–22 g	Varied	Choline HCl	Single dose	Oral	LD_{50}, 3,900 mg/kg LD_0, 2,000 mg/kg	Agut et al., 1983
Rats, male and female albinos, 413	76–343 g	300 mg/ml	Choline HCl	1 d	Stomach tube	LD_{50}, 3.4 g/kg	Neuman and Hodge, 1945
Rats, male and female albinos, 413	76–343 g	580 mg/ml	Choline HCl	1 d	Stomach tube	LD_{50}, 6.1 g/kg	Neuman and Hodge, 1945
Rats, albinos, 90	120–200 g	0.4–1.00 g	Choline HCl	1 d	Stomach tube	LD_{50}, 6.7 g/kg; liver and spleen appeared congested, stomach bleached and distended; blood vessels of stomach and diaphragm engorged	Hodge and Goldstein, 1942
Rats, 518		20–200 mg/ml	Choline HCl	Single injection	IP	At high levels, death within 20 min or not at all; at death, hemorrhage around eyes	Hodge, 1944
Rats, female Holtzmans, 50	52 g	1,646 mg/kg (choline)	Choline HCl	18 d	Diet	No adverse effects; liver lipid concentration reduced; growth improved when dietary methionine deficient	Kroenig and Pond, 1967
Rats, 5/group	55–60 g	0.01–100,000 mg/kg diet	Choline HCl	3–4 mo	Diet	Growth unaffected with up to 10,000 mg/kg, reduced with 50,000, and nil with 100,000 mg/kg diet; highest level resulted in increased organ weights	Hodge, 1945
Swine, Duroc-Jerseys, 12/group	2 d	500–2,000 mg/kg diet DM	Choline	8 wk	Diet	Growth improved with up to 1,000 mg/kg; no adverse effects	Neumann et al., 1949
Swine, 24	3 wk (5.3 kg)	1,646 mg/kg (choline) diet	Choline HCl	28 d	Diet	No adverse effects; growth improved when dietary methionine deficient	Kroenig and Pond, 1967
Swine, sows, 16–18/group	Mature	Dietary supplement of 0.64–1.5 g/kg diet	Choline HCl[a]	Weaning to farrowing	Diet	No reduction in incidence of splay leg in piglets; no adverse effects	Dobson, 1971

[a]It is presumed that this form was used.

on an accelerated program. It was dosed twice daily 7 days after the start, and then 3 times daily on the twenty-sixth day. Anemia was established in that animal more rapidly. Two other dogs were given doses of 10 mg/kg 3 times daily from the beginning of the experiment. As a result of 3 daily doses of choline hydrochloride, five dogs showed 30 to 43 percent reductions in red blood cell counts. Two other dogs exhibited milder anemias.

Hodge (1945) added choline chloride to the diet and drinking water of rats. He found that levels of up to 1 percent of choline chloride in the diet produced no evidence of toxicity. Growth was depressed at higher levels, however. When choline chloride was given in the drinking water, growth depression was observed at the 1 percent level. Levels greater than 3 percent were not well tolerated. In general, the effects of choline chloride on the histopathology of the major organs were negative.

Choline has been administered as a precursor of acetylcholine to humans suffering from various nervous and mental diseases. Hollister et al. (1978) reported that peak plasma concentration was obtained with repeated oral doses of 16 to 20 g/d, with a rapid disappearance following treatment cessation. Some clinical improvement and no deleterious effects were noted in those patients.

PRESUMED UPPER SAFE LEVELS

Insufficient data are available to support precise estimates of maximum tolerable dietary levels of choline, although published information suggests that the tolerance for choline is high in most species. For instance, McKibbin et al. (1944) supplemented the diet of growing pups with up to 1,500 mg of choline chloride/kg with no reported problems. McKibbin et al. (1945) used supplements of up to 2,000 mg/kg of diet successfully with growing pups.

Hodge (1944) found that the acute intraperitoneal LD_{50} values of choline chloride for rats ranged from 35 to 74 mg/100 g of BW. Lethality varied according to the concentration of the dosing solution. Hodge also noted that all deaths took place within 20 minutes after the injections. Hodge and Goldstein (1942) determined the LD_{50} values of choline chloride for both mice and rats. In mice injected intraperitoneally with an aqueous solution of 2 percent choline chloride, the LD_{50} was 320 mg/kg of BW. In rats given a 67 percent solution of the compound in the same manner, the LD_{50} was 6.7 g/kg of BW. Neuman and Hodge (1945) administered choline chloride in doses of four concentrations (200, 400, 500, and 670 mg/ml) to rats by stomach tube. The total lethality of the two higher concentrations was significantly

greater than that of the two lower concentrations. The pooled LD_{50} was 3.4 g/kg of BW for the higher concentration groups and 6.1 g/kg of BW for the lower concentration groups.

Studies with mice suggest that choline chloride is relatively innocuous when administered orally. According to Agut et al. (1983), the acute oral LD_{50} value of choline chloride is 3,900 mg/kg of BW. The acute intravenous LD_{50} is 53 mg/kg of BW. They estimated the oral and intravenous maximum tolerable levels to be 2,000 and 21.5 mg/kg of BW, respectively. The CDP form of choline was less toxic. The LD_{50} values were indeterminate for the oral route and 4,600 mg/kg of BW for the intravenous route. The oral and intravenous maximum tolerable levels were 14,000 mg/kg of BW and 3,500 mg/kg of BW, respectively. Wecker and Schmidt (1979) have shown that the CDP form is almost completely absorbed. Their work supports other data suggesting that the small amount of chloride in choline chloride may contribute to the toxicity of this compound.

SUMMARY

1. Although choline is not a vitamin in the strictest sense, it is an important nutrient as a source of labile methyl groups. Dietary requirements have been established for young chickens, swine, rats, and dogs for which endogenous synthesis is insufficient for physiological demands or is inadequate in circumstances of dietary deficiencies of methyl groups.

2. Data with pigs indicate a high tolerance for choline. Studies with chickens suggest that a dietary level of about twice the dietary requirement is safe and produces no deleterious effects. Some of the chicken data indicate a growth reduction and interference with the utilization of pyridoxine when the dietary level of choline exceeds twice the required level.

3. Studies with dogs suggest a low tolerance for choline chloride and lecithin in that species. Adverse effects have been reported for levels of choline chloride equivalent to 3 times the apparent choline requirement.

4. Mouse data suggest that choline chloride is relatively innocuous when given orally (the LD_{50} is 3,900 mg/kg of BW), but appreciably more toxic when given intravenously (the LD_{50} is 53 mg/kg of BW). The CDP form of choline is less toxic to mice by the same parameters; the oral LD_{50} is indeterminate and the intravenous LD_{50} is 4,600 mg/kg of BW. The maximum tolerable levels of choline chloride and CDP choline appear to be 2,000 and 14,000 mg/kg of BW, respectively, when given orally, and 21.5 and 3,500 mg/kg of BW, respectively, when given intraperitoneally.

5. The LD_{50} of choline chloride administered to rats by

stomach tube was estimated to be 3.4 to 6.1 g/kg of BW.

6. The fact that choline chloride appears to present some hazard to chickens and dogs when included in the diet at relatively low levels indicates a need for additional research on choline with these and other species.

REFERENCES

Agut, J., E. Font, A. Sacristan, and J. A. Ortiz. 1983. Dissimilar effects in acute toxicity studies of CDP-choline and choline. Arzneim-Forsch/Drug Res. 33:1016.

Beisel, W. R. 1982. Single nutrients and immunity. Am. J. Clin. Nutr. 35:417.

Berry, E. P., C. W. Carrick, R. E. Roberts, and S. M. Hauge. 1943. A deficiency of available choline in soybean oil and soybean oil meal. Poult. Sci. 22:442.

Chan, M. M. 1984. Choline and carnitine. P. 549 in Handbook of Vitamins, L. J. Machlin, ed. New York: Marcel Dekker.

Crawford, J. S., M. Griffith, R. A. Tuckell, and A. B. Watts. 1969. Choline requirement and synthesis in laying hens. Poult. Sci. 48:620.

Davis, J. E. 1944a. Depression of the normal erythrocyte number by soybean lecithin or choline. Am. J. Physiol. 142:65.

Davis, J. E. 1944b. The experimental production of a hyperchromic anemia in dogs which is responsive to anti-pernicious anemia treatment. Am. J. Physiol. 142:402.

Deeb, S. S., and P. A. Thornton. 1959. The choline requirement of the chick. Poult. Sci. 38:1198. (Abstr.)

Dobson, K. J. 1971. Failure of choline to prevent splay leg in piglets. Anat. Vet. J. 47:587.

Fritz, J. C., T. Roberts, and J. W. Boehne. 1967. The chick's response to choline and its application to an assay for choline in feedstuffs. Poult. Sci. 46:1447.

Hodge, H. C. 1944. Acute toxicity of choline hydrochloride administered intraperitoneally to rats. Proc. Soc. Exp. Biol. Med. 57:26.

Hodge, H. C. 1945. Chronic oral toxicity of choline chloride in rats. Proc. Soc. Exp. Biol. Med. 58:212.

Hodge, H. C., and M. R. Goldstein. 1942. The acute toxicity of choline hydrochloride in mice and rats. Proc. Soc. Exp. Biol. Med. 51:281.

Hollister, L. E., D. J. Jenden, J. R. D. Amaral, J. D. Barchas, K. L. Davis, and P. A. Berger. 1978. Plasma concentrations of choline in man following choline chloride. Life Sci. 23:17.

Jukes, T. H. 1941. Studies of perosis in turkeys. 1. Experiments related to choline. Poult. Sci. 20:251.

Ketola, H. G., and M. C. Nesheim. 1974. Influence of dietary choline and methionine levels on the requirement for choline by chickens. J. Nutr. 101:1404.

Kroenig, G. H., and W. G. Pond. 1967. Methionine, choline and threonine interrelationships for growth and lipotropic action in the baby pig and rat. J. Anim. Sci. 26:352.

March, B. E. 1981. Choline supplementation of layer diets containing soybean meal or rapeseed meal as protein supplement. Poult. Sci. 60:818.

March, B. E., and C. MacMillan. 1979. Trimethylamine production in the caeca and small intestine as a cause of fishy taints in eggs. Poult. Sci. 58:93.

Marvel, J. A., C. W. Carrick, R. E. Roberts, and S. M. Hauge. 1943. The supplementary value of choline and methionine in a corn and soybean oil meal chick ration. Poult. Sci. 23:294.

McKibbin, J. M., S. Thayer, and F. J. Stare. 1944. Choline deficiency studies in dogs. J. Lab. Clin. Med. 29:1109.

McKibbin, J. M., R. M. Ferry, Jr., S. Thayer, E. G. Patterson, and F. J. Stare. 1945. Further studies on choline deficiency in dogs. J. Lab. Clin. Med. 30:422.

Mishler, D. H., C. W. Carrick, R. E. Roberts, and S. M. Hauge. 1946. Synthetic and natural vitamin supplements for corn and soybean oil meal chick rations. Poult. Sci. 25:479.

Molitoris, B. A., and D. H. Baker. 1976. Assessment of the quantity of biologically available choline in soybean meal. J. Anim. Sci. 42:481.

Mookerjea, S. 1971. Action of choline in lipoprotein metabolism. Fed. Proc. 30:143.

National Research Council. 1979. Nutrient Requirements of Swine, 8th rev. ed. Washington, D.C.: National Academy Press.

National Research Council. 1984. Nutrient Requirements of Poultry. 8th rev. ed. Washington, D.C.: National Academy Press.

Nauss, K. M., and P. M. Newberne. 1980. Effects of dietary folate, vitamin B12 and methionine/choline deficiency on immune function. Adv. Exp. Med. Biol. 135:63.

Neuman, M. W., and H. C. Hodge. 1945. Acute toxicity of choline chloride administered orally to rats. Proc. Soc. Exp. Biol. Med. 58:87.

Neumann, A. L., J. L. Krider, M. F. James, and B. C. Johnson. 1949. The choline requirement of the baby pig. J. Nutr. 38:195.

Saville, D. G., A. Solvyns, and C. Humphries. 1967. Choline-induced pyridoxine deficiency in broiler chickens. Aust. Vet. J. 43:346.

Wecker, L., and D. E. Schmidt. 1979. Central cholinergic function: Relationships to choline administration. Life Sci. 25:375.

Research Needs

To make precise quantitative estimates of the vitamin tolerances of animals, one needs a reasonably complete base of experimental and clinical information. The available information base is presently insufficient to support such estimates, however. The toxicities of some vitamins, such as vitamin K, folic acid, pantothenic acid, and vitamin B_{12}, are so low that hypervitaminosis is very unlikely in practical situations of animal feeding. The lack of more extensive data is, therefore, without consequence. For other vitamins that have measurable toxicities and may be used in practical situations at levels greater than nutritional requirements, however, the gap in the information is much more serious. The review of the current vitamin literature points to several such gaps. In order to fill them, research should be initiated in the following areas.

1. Vitamin A: The clear potential for hypervitaminosis A and the small amount of quantitative data describing its safe dietary levels for most species of animals indicates a need to further define those levels, particularly for chronic exposure of domestic species. At a more fundamental level, the biochemical mechanisms of vitamin A toxicity need to be further elucidated.

2. Vitamin D: Further quantitative information is needed to define more clearly dietary levels of vitamin D that are safe for domestic animal species. Studies are also needed to make clear hypercalcemia's role in tissue calcinosis due to hypervitaminosis D. Because of the physiological factors that affect and interspecific differences in circulating levels of 25-hydroxycholecalciferol, the usefulness of this parameter as a clinical indicator of hypervitaminosis D needs to be evaluated.

3. Vitamins E and C: Because of the increasing use of high levels of these vitamins as promoters of immune functions and protectors against stress, the chronic toxicities of each should be more carefully defined for a variety of domestic animal species.

4. Niacin: The effects of excesses of nicotinamide and nicotinic acid on hepatic and renal function should be elucidated.

5. Choline: The relative toxicities of high levels of the different chemical forms of choline should be re-evaluated in a manner that considers the potential effects of the chloride provided by choline chloride.

Summary

The scientific literature provides information concerning the adverse effects of high-level exposures to most of the vitamins. Unfortunately, this information base is limited with respect to the species investigated, the forms of vitamins studied, and the types of experimental design used. Therefore, the maximum tolerable levels of animals for several of the vitamins were estimated only from a limited base of information.

It is apparent that toxicity in animals is greatest for high-level oral exposures to vitamin A, vitamin D, and choline (in the form of choline chloride). For each, about 10-fold the dietary level required to prevent deficiency diseases can depress growth and adversely affect particular organ systems. Signs of hypervitaminosis A can be observed among animals fed 10 to 30 times the vitamin levels required to prevent deficiency diseases. Hypervitaminosis A is characterized by disturbances of nervous function, such as hyperirritabiity, twitching, convulsions, and paralysis. It is also characterized by liver dysfunction and skin disorders. Nonruminants appear to be able to tolerate dietary levels of at least 30-fold their vitamin A requirements. Ruminants may be able to tolerate only 10-fold their required levels. Signs of hypervitaminosis D can be observed among animals fed 4- to 10-fold their nutritional requirements for more than 60 days. For less than 60 days, however, animals can tolerate as much as 1,000-fold the required levels without adverse effects. Hypervitaminosis D is characterized by anorexia, gastrointestinal distress, lameness, polyuria, hypercalcemia, and calcinosis (particularly cardiovascular and renal calcification). It is not known whether calcinosis involves specific tissue lesions induced by high levels of the vitamin, or whether it is simply a consequence of the induced hypercalcemia. Depressed growth has been observed in response to dietary supplements of choline at 2 to 4 times the requirements. Most, if not all, of this toxicity may be due to disturbances in acid-base relations in the use of the chloride salt of this vitamin, however.

Niacin, riboflavin, and pantothenic acid are generally tolerated by animals at dietary levels as great as 10- to 20-fold their respective nutritional requirements. Niacin hypervitaminosis is characterized by reduced growth and by disturbances in the metabolism of lipoproteins and foreign compounds. Niacin can also affect nervous function. Nicotinic acid appears to be less toxic than nicotinamide. The pathology of riboflavin hypervitaminosis is not well described. It is clear, however, that its toxicity is greater when administered parenterally than when administered orally. High levels of pantothenic acid can cause liver damage characterized by hepatic steatitis and elevated serum transaminase activities.

Vitamin E is generally tolerated at dietary intakes as great as 100-fold nutritional required levels of animals. Hypervitaminosis E is characterized by reduced growth, reduced hematocrit, reticulocytosis, hepatic dysfunction, and hypoprothrombinemia. The last effect is apparently due to an antagonism of the utilization of vitamin K.

Vitamins K and C, thiamin, and folic acid are generally tolerated at oral intake levels of at least 1,000-fold animals' respective nutritional requirements. Menadione has been shown to be nephrotoxic; however, other forms of vitamin K are essentially innocuous. In laboratory animals, high levels of ascorbic acid (vitamin C) can produce oxaluria, uricosuria, hypoglycemia, excessive absorption of iron, gastrointestinal disturbances, allergic reactions, and anemia. These effects have not been extensively studied in domestic animals, few of which require this factor in their diets. Thiamin hypervitaminosis is characterized by impaired nervous function (for example, epileptiform convulsions and respiratory paralysis). The mechanism by which high levels of thiamin interfere with nervous function remains to be made clear. Pyridoxine excess can result in ataxia and peripheral neuropathy involving demyelination of nervous tissue. This condition has only been studied in dogs and rats. Folic acid hypervitaminosis due to oral exposure has not been reported. Parenteral administration of massive amounts of the vitamin has been found to produce epileptiform convulsion and renal hypertrophy in rats, however.

Very little information is available concerning the pathologies of biotin and vitamin B_{12} hypervitaminoses. Poultry and swine can easily tolerate biotin dietary levels 4 times their nutritional requirements. It is probable that much greater levels may be tolerated. The present scientific literature is unclear and most incomplete concerning animal tolerance of vitamin B_{12}. It is likely, however, that even high levels of this vitamin are essentially innocuous.

Appendix Table

APPENDIX TABLE   Estimated Vitamin Requirements of Domestic and Laboratory Animals (Dry Diet Bases)

Species	Vitamin A (IU/kg BW/day)	Vitamin A (IU/kg diet)	Vitamin D (IU/kg BW/day)	Vitamin D (IU/kg diet)	Vitamin E (IU/kg diet)	Vitamin K (menadione) (mg/kg diet)	Vitamin C (mg/kg diet)
Birds[a]							
Chickens							q
growing chicks		1,500		200	10	0.5	
laying hens		4,000		500	5	0.5	
breeding hens		4,000		500	10	0.5	
Ducks							q
starting and growing ducks		4,000		220		0.4	
breeding ducks		4,000		500		0.4	
Geese							q
starting and growing geese		1,500		200			
breeding geese		4,000		200			
Pheasants							q
starting and growing pheasants							
Quail							q
starting bobwhite quail							
breeding bobwhite quail							
starting and growing Japanese quail		5,000		1,200	12	1	
breeding Japanese quail		5,000		1,200	25	1	
Turkeys							q
growing poults		4,000		900	10–12	0.8–1.0	
breeding turkey hens		4,000		900	25	1	
Cats[b]		10,000		1,000	80		q
Cattle							q
dry heifers[c]	42	2,200	6.6	300			
dairy bulls[c]	42	2,200	6.6	300			
lactating cows[c]	76	3,200		300			
pregnant dry cows[c]		3,200		300			
beef cattle[d]		2,200		275			
Dogs[e]		5,000		500	50		q
Fishes							
bream[f]							
carp[f]		10,000			200–300		
catfish[f]		1,000–2,000		500–1,000	30		60
coldwater species[g]		2,500		2,400	30	10	100
Foxes[h]		2,440					q
Goats[i]	24–60		4.8–12.9				q
Guinea pigs[j]		23,333		1,000	50	5	200
Hamsters[k]		3,636		2,484	3	4	q
Horses[l]							q
ponies	25						
gestating mares	50						
lactating mares	55–65						
weanlings, yearlings	40						
2-year-olds	30						
Mice[k]		500		150	20	3	q
Mink[h]		5,930			27		q
Nonhuman primates[m]		10,000–15,000		2,000	50		100
Rabbits[n]							q
growing		580			40		
pregnant		>1,160			40	0.2	
lactating					40		
Rats[k]		4,000		1,000	30	0.5	q

Thiamin (mg/kg diet)	Niacin (mg/kg diet)	Riboflavin (mg/kg diet)	Pyridoxine (mg/kg diet)	Folacin (mg/kg diet)	Pantothenic Acid (mg/kg diet)	Biotin (mg/kg diet)	Vitamin B_{12} (μg/kg diet)	Choline (mg/kg diet)
1.8	27	3.6	2.5–3.0	0.25–0.55	10	0.1–0.15	3–9	500–1,300
0.8	10	2.2	3	0.25	2.2	0.1	4	
0.8	10	3.8	4.5	0.35	10	0.15	4	
	55	4	2.6		11			
	40	4	3		11			
	35–55	2.5–4.0			15			
	20	4						
	40–60	3.0–3.5			10			1,000–1,500
	30	3.8			13			1,500
	20	4			15			1,000
2	40	4	3	1	10	0.3	3	2,000
2	20	4	3	1	15	0.15	3	1,500
2.0	40–70	2.5–3.6	3.0–4.5	0.7–1.0	9–11	0.1–0.2	3	800–1,900
2.0	30	4	4	1	16	0.15	3	1,000
5	45	5	4	1	10	0.5	20	2,000
1	11.4	2.2	1	0.18	10	0.1	22	1,250
			5–6		30–50	1		4,000
	28	7	5–6		10–20			
1	14	9	3	5	40	1	20	3,000
10	150	20	10					
1	9.6	3.7–5.5	1.8	0.2	7.4			
2	10	3	3	4	20	0.3	10	1,000
20	90	15	6	2	40	0.6	10	2,000
5	10	7	1	0.5	10	0.2	10	600
1.3	20	1.6	1.6	0.5	8	0.12	32.6	
	50	5	2.5	0.2	15	0.1		
	180		39					1,200
4	20	3	6	1	8		50	1,000

APPENDIX TABLE—*Continued*

Species	Vitamin A (IU/kg BW/day)	Vitamin A (IU/kg diet)	Vitamin D (IU/kg BW/day)	Vitamin D (IU/kg diet)	Vitamin E (IU/kg diet)	Vitamin K (menadione) (mg/kg diet)	Vitamin C (mg/kg diet)
Sheep[o]							[q]
ewes, nonlactating, early gestation	26		5.6				
ewes, late gestation, lactation	35		5.6				
rams	43		5.6				
lambs, early-weaned	35		6.6				
lambs, finishing	26		5.5				
Shrimps[f]							10,000
Swine[b]							[q]
growing, finishing		1,300–2,200		125–200	11	2	
bred gilts and sows, boars		4,000		200	10	2	
lactating gilts and sows		2,000		200	10	2	

[a]National Research Council. 1984. Nutrient Requirements of Poultry. 8th rev. ed. Washington, D.C.: National Academy Press. 71 pp.

[b]National Research Council. 1978. Nutrient Requirements of Cats. Rev. ed. Washington, D.C.: National Academy Press. 56 pp.

[c]National Research Council. 1978. Nutrient Requirements of Dairy Cattle. 5th rev. ed. Washington, D.C.: National Academy Press. 76 pp.

[d]National Research Council. 1984. Nutrient Requirements of Beef Cattle. 6th rev. ed. Washington, D.C.: National Academy Press. 90 pp.

[e]National Research Council. 1985. Nutrient Requirements of Dogs. Rev. ed. Washington, D.C.: National Academy Press. 56 pp.

[f]National Research Council. 1983. Nutrient Requirements of Warmwater Fishes and Shellfishes. Rev. ed. Washington, D.C.: National Academy Press. 63 pp.

[g]National Research Council. 1981. Nutrient Requirements of Coldwater Fishes. Washington, D.C.: National Academy Press. 63 pp.

[h]National Research Council. Nutrient Requirements of Mink and Foxes. 1982. 2nd rev. ed. Washington, D.C.: National Academy Press. 72 pp.

[i]National Research Council. 1981. Nutrient Requirements of Goats: Angora, Dairy, and Meat Goats in Temperate and Tropical Countries. Washington, D.C.: National Academy Press. 91 pp.

[j]Recommended dietary allowances (not requirements) are given for this species.

[k]National Research Council. 1978. Nutrient Requirements of Laboratory Animals. 3rd rev. ed. Washington, D.C.: National Academy Press. 96 pp.

[l]National Research Council. 1978. Nutrient Requirements of Horses. 4th rev. ed. Washington, D.C.: National Academy Press. 33 pp.

[m]National Research Council. Nutrient Requirements of Nonhuman Primates. 1978. Washington, D.C.: National Academy Press. 83 pp.

[n]National Research Council. 1977. Nutrient Requirements of Rabbits. 2nd rev. ed. Washington, D.C.: National Academy Press. 33 pp.

[o]National Research Council. 1975. Nutrient Requirements of Sheep. 5th rev. ed. Washington, D.C.: National Academy Press. 72 pp.

[p]National Research Council. 1979. Nutrient Requirements of Swine. 8th rev. ed. Washington, D.C.: National Academy Press. 52 pp.

[q]Ascorbic acid is not a dietary essential for this species.

Thiamin (mg/kg diet)	Niacin (mg/kg diet)	Riboflavin (mg/kg diet)	Pyridoxine (mg/kg diet)	Folacin (mg/kg diet)	Pantothenic Acid (mg/kg diet)	Biotin (mg/kg diet)	Vitamin B_{12} (μg/kg diet)	Choline (mg/kg diet)
120			120	120				600
1.1–1.3	10–22	2.2–3.0	1.1–1.5	0.6	11–13	0.1	11–22	400–1,100
	10	3	1	0.6	12	0.1	15	1,250
	10	3	1	0.6	12	0.1	15	1,250

Index